A Prelude to Quantum Field Theory

A Prelude to Quantum Field Theory

JOHN DONOGHUE
LORENZO SORBO

PRINCETON UNIVERSITY PRESS
Princeton and Oxford

Published by Princeton University Press
41 William Street, Princeton, New Jersey 08540
6 Oxford Street, Woodstock, Oxfordshire OX20 1TR

press.princeton.edu

All Rights Reserved

Library of Congress Cataloging-in-Publication Data

Names: Donoghue, John, 1950– author. | Sorbo, Lorenzo, 1973– author.
Title: A prelude to quantum field theory / John Donoghue and Lorenzo Sorbo.

Description: Princeton : Princeton University Press, [2022] | Includes bibliographical references and index.
Identifiers: LCCN 2021034292 (print) | LCCN 2021034293 (ebook) | ISBN 9780691223490 (hardback) | ISBN 9780691223483 (paperback) | ISBN 9780691223506 (ebook)
Subjects: LCSH: Quantum field theory. | BISAC: SCIENCE / Physics / Quantum Theory
Classification: LCC QC174.45 .D66 2022 (print) | LCC QC174.45 (ebook) | DDC 530.14/3—dc23
LC record available at https://lccn.loc.gov/2021034292
LC ebook record available at https://lccn.loc.gov/2021034293

British Library Cataloging-in-Publication Data is available

Editorial: Ingrid Gnerlich, Whitney Rauenhorst
Jacket/Cover Design: Wanda España
Production: Danielle Amatucci
Publicity: Matthew Taylor, Charlotte Coyne
Copyeditor: Karen B. Hallman

Jacket/Cover Credit: local_doctor / Shutterstock

This book has been composed in Minion Pro & Universe

Printed on acid-free paper ∞

Printed in the United States of America

10 9 8 7 6 5 4 3 2 1

Contents

Preface

Quantum Field Theory is the ultimate way to understand quantum physics. It is an incredibly beautiful subject, once you get used to the field-theoretical way of thinking. After teaching it for many years, we have found that the primary hurdle is to make the transition from the way a student thinks of quantum mechanics to the way we think in field theory. Once that has been accomplished there are many fine books on Quantum Field Theory that can guide you further into this rich topic.

This book is dedicated to helping students make this transition. It is not a complete exposition of Quantum Field Theory. There are many books that cover Quantum Field Theory in much greater detail. However, our experience has been that many students struggle at the start of this transition—for several reasons. One is certainly connected to their quantum-mechanical backgrounds. The styles of thinking in quantum mechanics and Quantum Field Theory are different and not always adequately addressed in the classic texts. Another aspect is that Quantum Field Theory is an elegant and very extensive branch of physics, and the classic textbooks tend to be elegant and very large, which are not the best choices for a novice. Many books tend to be either particle-centric or condensed matter–centric and are difficult for a mixed audience in the classroom. And many modern texts start with path integrals. For the science, this is logical because path integrals are a deep way to understand quantum theory. However, it is a difficult starting point for a student who is just emerging from the Schrödinger equation.

Another motivation for this book is to provide insight into Quantum Field Theory for those who do not intend to practice it personally or who are not sure about the direction of their future studies. This could be general readers who have developed an understanding of quantum mechanics and who want to learn a bit more about the field-theoretical version of quantum physics. Advanced undergraduate students often want to learn what Quantum Field Theory is about, and we hope that they find this book approachable. Among professional physicists, there are many whose research does not involve Quantum Field Theory directly but who wish to understand its language and methods. One does not have to digest a large comprehensive volume on Quantum Field Theory to develop an appreciation for the subject. This book provides the entrée to the field, and for some readers this could be sufficient.

This is meant to be a modest book designed to fill these gaps in the pedagogic literature. We take a quantum-mechanical starting point and change it into a field theory version. We strive to emphasize at all stages the philosophy behind the study of field theory. Whenever possible, connections are made (stressing both similarities and differences) to the older style of thinking. We use canonical quantization at the start, because we have found that it is most useful for the initial discussion of the free particle states of the theory. Then we repeat the process using the path integral language to show how that method produces the same result and to introduce the reader to that style of treatment of Quantum Field Theory. The discussion will involve primarily scalar fields to avoid the complications of Dirac and gauge fields. We do include some discussion of fermions, but the main development of the subject does not rely on this material.

We intend for this book to be subfield-neutral; that is, appropriate for physicists from any of the subdisciplines or for learned nonphysicists. However, we do choose to use relativistic notation (after introducing it) because it is notationally clean, allows the important physics to be seen more clearly, and in the end matches the notation of most other books. For those planning further study of Quantum Field Theory, we suggest following this book with either a standard field theory book or a condensed matter book. The final chapter is a guide to "the rest of the story," which is concise, without derivations, but it is intended to provide a guide to the next steps on the journey through Quantum Field Theory.

The book could be used in a one-semester Quantum Field Theory class at either the undergraduate or graduate levels. Undergraduate instructors could tailor the presentation to match the preparation of their students. The book covers the material that we include in our Quantum Field Theory I courses when we have a mix of students from the particle/nuclear and condensed matter research areas. One model that we would favor is to have *all* graduate physics students take a course like ours, with subfield specialization only occurring afterward.

We would like to thank our colleagues for their many insights into Quantum Field Theory over the years. They are so many that it would be impossible to list all their names here. We would also like to thank our Physics 811 students, who helped us shape this book with their questions and comments. We additionally gratefully acknowledge our funding from the U.S. National Science Foundation.

Princeton University Press and the authors will maintain a web page with potential errata, supplemental materials, problems, and exercises. This can be found at https://press.princeton.edu/books/a-prelude-to-quantum-field-theory and at https://blogs.umass.edu/preludeqft/. This material can also be accessed through the professional home pages of the authors.

CHAPTER 1

Why Quantum Field Theory?

Quantum theory began with quanta of the electromagnetic field, with Planck's blackbody spectrum and with Einstein's concept of a photon of energy $E = \hbar\omega$ in the photoelectric effect. The latter process is driven by a photon in the initial state that is absorbed and does not appear in the final state. However, the standard curriculum of quantum mechanics initially bypasses this topic and instead emphasizes wavefunctions and quantum properties of massive particles at low energies, such as those typical of atomic physics. In these cases, the particle number is conserved. To provide a proper quantum treatment of the emission and absorption of the electromagnetic quanta, one needs Quantum Field Theory. We need to transition from thinking about wavefunctions to discussing fields. As we will see, this allows the numbers and identities of the particles to change in reactions, which has wide applicability. In fact, this is a unifying concept. Not only does the electromagnetic field behave like a particle in certain settings, but also the entities that we think of as *particles*, such as electrons, can behave like *waves* in diffraction experiments. Moreover, all types of particles can be created and destroyed, as in the reaction $e^+ e^- \rightarrow 2\gamma$ with e^- being the electron, e^+ the positron, and γ being the photon. To describe such processes, we need Quantum Field Theory.

Another indication of the need for Quantum Field Theory arose in attempts to marry quantum mechanics and the Theory of Special Relativity. Schrödinger's first attempts to write a differential equation whose solutions would describe de Broglie's matter waves were based on applying the identification $E \leftrightarrow i\hbar\partial_t$, $\mathbf{p} \leftrightarrow -i\hbar\nabla$ to the relativistic energy-momentum relation $E^2 = m^2 c^4 + \mathbf{p}^2 c^2$ for a particle of mass m (an effort that would eventually lead to the Klein-Gordon equation). This construction, however, led to complications (negative probability densities as well as negative energy states), and Schrödinger went ahead with the more modest(!) goal of writing a nonrelativistic equation for the hydrogen atom. Nonrelativistic quantum mechanics was thus born as a plan B, because Schrödinger noticed that its relativistic counterpart was leading to mathematical and physical inconsistencies. Indeed, it took another couple of decades or so to realize that a consistent treatment of relativistic quantum mechanics requires a quantum theory of fields and to formalize this theory. Along the way, people had to deal with the subtleties that we will discuss later in this book. These involved abandoning the concept of wavefunction in favor of a field *operator* acting on a Hilbert space of states.

1.1 A successful framework

Quantum Field Theory has been successful in making predictions throughout all branches of quantum physics. While the framework of this theory was developed to describe electrons and photons, it finds applications in the theory of elementary particles, in macroscopic systems of condensed matter physics, and in the Early Universe. Today, it is impossible to list all the successful applications of the quantum theory of fields. Based on the highly arbitrary choices of the authors, these include:

- one of the most iconic predictions of Quantum Field Theory: the existence of antimatter, which emerged from the formulation of Dirac's equation. Dirac was motivated by the need of making sense of the relativistic relation $E^2 = m^2 c^4 + \mathbf{p}^2 c^2$ without incurring the presence of negative norm states. A major experimental fact (the prediction of a new form of matter, antimatter, four years before the discovery of the positron) stemmed from a strictly mathematical requirement. With that same equation, Dirac also postdicted the ratio (in appropriate units approximately equal to 2) between the magnetic moment of the electron and its spin (the so-called gyromagnetic ratio or *g*-factor of the electron), which was not justified by any existing theory at that time.
- the exquisite agreement of the value of the electron's *g*-factor as predicted by Quantum Field Theory with its measured value is one of the quantities most precisely measured in physics, where the quantity $(g-2)/2$ is measured[1] to be 0.0115965218073(28) in very close agreement with the theoretical value 0.0115965218161(23) when using the most precise direct measurement of the fine structure constant. The corrections to the Dirac value $g = 2$ come from loop diagrams, which we will discuss starting in chapter 5.
- the prediction of the Lamb shift—the energy difference between the $^2S_{1/2}$ and $^2P_{1/2}$ energy levels of hydrogen. The levels are degenerate even in the relativistic Dirac theory. A full treatment required the development of Quantum Electrodynamics.
- Landau-Ginzburg's theory of superconductivity, based on Landau's theory of phase transitions, in which the behavior of macroscopic systems is described after coarse-graining their microscopic component.
- the running of coupling constants. In physical processes, the values of the coupling constants depend on the energy or distance scale at which they are measured. This effect is observed both in particle physics and in condensed matter systems near phase transitions, where the running affects the

---

[1]This value, plus reviews of many tests of Quantum Field Theory predictions, can be found in the *Review of Particle Properties*, which is maintained and updated regularly by P. A. Zyla et al. for the Particle Data Group. See also Particle Data Group et al., "Review of Particle Physics," *Progress of Theoretical and Experimental Physics* 2020, no. 8 (August 2020): 083C01.

value of the *critical exponents*, that is, the way certain quantities evolve as we change the temperature of the system near a phase transition.

- the origin of structure in the Early Universe. While on large scales the Universe is largely uniform, on small scales we see clumping of matter as well as voids. There are strong indications that this is the result of the amplification of quantum fluctuations in the Early Universe, which can be described by Quantum Field Theory.

1.2 A universal framework

The point of view that informs this book is that the main reason why we need Quantum Field Theory is because it is *universal*. As we will see in chapter 2, any system governed by quantum mechanics in which we ignore the ultimate microscopic behavior is controlled, at sufficiently low energies/long wavelengths, by the rules of Quantum Field Theory. This is true for the description of the sound waves that propagate in your desk when you hit it. It could even be true for our "elementary" particles, as our understanding of what is elementary has changed over time. (For example, the proton and neutron were once considered elementary, but now we have a more fundamental description in terms of quarks and gluons.) At whatever scale we are working, Quantum Field Theory can be an appropriate description.

In the end, Quantum Field Theory provides an elegant and understandable treatment of *all* particles. All fields are treated on the same footing. Its rules theory can handle all types of transitions with powerful techniques. Quantum Field Theory is conceptually unified and clear once one learns how to think appropriately about the subject.

CHAPTER 2

Quanta

To start down the path to Quantum Field Theory, we have to first head back to 1905, when Einstein first postulated that photons carry energy in quanta of $\hbar\omega$, and uncover the quantum of a field. The same transition from classical physics to quantum physics that works in ordinary quantum mechanics will also work here. This leads to the quantization rules for fields and to the concepts of field operators and particle quanta.

2.1 From classical particle mechanics to classical waves: Phonons

Mathematically, fields are functions of space and time, such as a function $\phi(t, x)$. For us as physicists, this also means that they satisfy wave equations, that they carry energy, that they ultimately have interactions, etc. Let us start by constructing a field in a way that also allows us to quantize it.

Consider a one-dimensional array of particles of mass m with coordinates $y_j(t)$ interacting with their neighbors, as in figure 2.1. Near the equilibrium configuration, the potential can be approximated by a harmonic oscillator, that is, by a set of springs. We will denote by a the rest length of the springs and by k the spring constant, so that the interaction term between the $(j+1)$-th and the j-th particle is $\frac{k}{2}(y_{j+1} - y_j - a)^2$. Denoting $\delta y_j(t) \equiv y_j(t) - aj$ as the deviation of the j-th particle from its equilibrium position, the system is described by a Lagrangian

$$L(\delta y_j, \delta \dot{y}_j) = \sum_j \left[\frac{m}{2} \delta \dot{y}_j^2 - \frac{k}{2} (\delta y_{j+1} - \delta y_j)^2 \right].$$ (2.1)

(We will assume here that the string is infinitely long, so that we do not have to worry about boundary conditions.) Physically we know that if the spacing is very small compared to the wavelength, $a \ll \lambda$, the system will be described by wavelike solutions, like sound waves propagating through a solid. When quantized this becomes a one-dimensional model for phonons in a solid.

The techniques of Lagrangian mechanics instruct us to define a canonical momentum

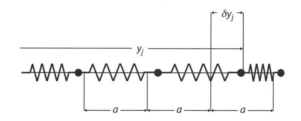

Figure 2.1. The system with many particles that provides our starting point. The Lagrangian for this system is given in equation (2.1).

$$\delta p_j \equiv \frac{\partial \mathcal{L}}{\partial(\delta \dot{y}_j)} = m\,\delta \dot{y}_j \tag{2.2}$$

and a Hamiltonian

$$H = \sum_j \delta p_j\,\delta \dot{y}_j - L(\delta y_j, \delta \dot{y}_j) = \sum_j \left[\frac{\delta p_j^2}{2m} + \frac{k}{2}(\delta y_{j+1} - \delta y_j)^2 \right]. \tag{2.3}$$

Now let us look at this over such large distances that the continuum limit is a good approximation. Mathematically, we obtain this by sending the distance $a \to 0$ and the number of sites to infinity, so that $a \times j$ stays finite. We will thus describe the position by a continuous variable

$$x = aj, \quad \text{so that} \quad \sum_j \to \int dj = \int \frac{dx}{a}. \tag{2.4}$$

It then makes sense to describe the displacement $\delta y_j(t)$ by a continuous function $\phi(t,x)$

$$\delta y_j(t) = \sqrt{\frac{1}{ka}}\,\phi(t,x). \tag{2.5}$$

The normalization constant shown does not change the physics in the end,[1] but is chosen to make the intermediate steps look cleaner. To proceed, we note that in the continuum limit $\delta y_{j+1}(t) - \delta y_j(t) \to a\,\partial_x \phi(t,x)/\sqrt{ka}$. We are now in position to derive the expression of the field-theoretical Lagrangian. The "potential" term takes the form

$$\sum_j \frac{k}{2}(\delta y_{j+1} - \delta y_j)^2 \to \int \frac{dx}{a} \times \frac{k}{2} \times \left(\frac{a}{\sqrt{ka}}\frac{\partial \phi}{\partial x} \right)^2 = \int dx\,\frac{1}{2}\left(\frac{\partial \phi}{\partial x} \right)^2, \tag{2.6}$$

while the kinetic energy term reads

---

[1] To verify this statement, one has to work through the steps leading up to the quantized Hamiltonian, equation (2.18). The equations of motion will be unchanged, and the canonical momentum will have a compensating change in normalization.

$$\sum_j \frac{m}{2} \delta \dot{y}_j^2 \rightarrow \int \frac{dx}{a} \times \frac{m}{2} \times \frac{1}{ka} \left(\frac{\partial \phi}{\partial t} \right)^2 \equiv \int dx \frac{1}{2v^2} \left(\frac{\partial \phi}{\partial t} \right)^2 , \qquad (2.7)$$

where we have defined the quantity

$$v = \sqrt{\frac{k}{m}} \, a , \qquad (2.8)$$

that has the dimensions of a velocity.

The end result is an action

$$S = \int dt \, dx \, \mathcal{L}(\partial_t \phi, \partial_x \phi) = \int dt \, dx \left[\frac{1}{2v^2} \left(\frac{\partial \phi}{\partial t} \right)^2 - \frac{1}{2} \left(\frac{\partial \phi}{\partial x} \right)^2 \right] , \qquad (2.9)$$

so that the action is written as the integral over the entire *spacetime* of a *Lagrangian density* $\mathcal{L}$.

We can derive the wave equation either by varying the original Lagrangian and taking the continuum limit or by directly varying the continuum action. To do the latter, we define the small variation $\delta \phi$

$$\phi(t, x) = \bar{\phi}(t, x) + \delta \phi(t, x) , \qquad (2.10)$$

with the variation vanishing at the endpoints and set the first variation of the action to 0

$$\delta S = 0 = \int dt \, dx \left[\frac{1}{v^2} \frac{\partial \bar{\phi}}{\partial t} \frac{\partial \delta \phi}{\partial t} - \frac{\partial \bar{\phi}}{\partial x} \frac{\partial \delta \phi}{\partial x} \right]$$

$$= \int dt \, dx \left[-\frac{1}{v^2} \frac{\partial^2 \bar{\phi}}{\partial t^2} + \frac{\partial^2 \bar{\phi}}{\partial x^2} \right] \delta \phi(t, x) . \qquad (2.11)$$

The second line is obtained by integrating by parts, with the surface term vanishing because we required $\delta \phi$ to vanish at the boundaries of the system. By requiring that the variation of the action vanishes for any $\delta \phi(t, x)$, we get the wave equation

$$\left[\frac{1}{v^2} \frac{\partial^2}{\partial t^2} - \frac{\partial^2}{\partial x^2} \right] \bar{\phi}(t, x) = 0 . \qquad (2.12)$$

This is the *Euler-Lagrange equation of motion* for this field.

The canonical momentum for the field ϕ can also be constructed in analogy with the usual coordinate construction

$$\delta p_j \equiv \frac{\partial \mathcal{L}}{\partial (\delta \dot{y}_j)} \Rightarrow \pi(t, x) = \frac{\partial \mathcal{L}}{\partial \dot{\phi}} = \frac{1}{v^2} \frac{\partial \phi}{\partial t} . \qquad (2.13)$$

For the present calculation, it is useful to display the exact relation

$$\delta p_j \equiv \frac{\partial \mathcal{L}}{\partial(\delta \dot{y}_j)} = m \, \delta \dot{y}_j = \frac{m}{\sqrt{ka}} \dot{\phi} = \frac{a\sqrt{ka}}{v^2} \dot{\phi} = a\sqrt{ka}\,\pi(t,x) \,. \tag{2.14}$$

This, in particular, tells us that

$$\sum_j \delta p_j \, \delta \dot{y}_j = \int \frac{dx}{a} \times \frac{m}{\sqrt{ka}} \dot{\phi} \times \frac{1}{\sqrt{ka}} \dot{\phi} = \int dx \, \frac{1}{v^2} \left(\frac{\partial \phi}{\partial t} \right)^2 = \int dx \, \pi(t,x) \, \dot{\phi}(t,x) \,. \tag{2.15}$$

The Hamiltonian is then easy to construct either through the continuum limit

$$H = \sum_j \left[\frac{\delta p_j^2}{2m} + \frac{k}{2} (\delta y_{j+1} - \delta y_j)^2 \right] = \int dx \left[\frac{1}{2v^2} \left(\frac{\partial \phi}{\partial t} \right)^2 + \frac{1}{2} \left(\frac{\partial \phi}{\partial x} \right)^2 \right] \,, \tag{2.16}$$

or through the field-theoretical *Hamiltonian density* $\mathcal{H}$

$$H = \int dx \, \mathcal{H} \tag{2.17}$$

defined by

$$\mathcal{H} = \pi \, \dot{\phi} - \mathcal{L} \,, \tag{2.18}$$

which in this case is equal to

$$\mathcal{H} = \frac{v^2}{2} \pi^2 + \frac{1}{2} \left(\frac{\partial \phi}{\partial x} \right)^2 = \frac{1}{2v^2} \left(\frac{\partial \phi}{\partial t} \right)^2 + \frac{1}{2} \left(\frac{\partial \phi}{\partial x} \right)^2 \,. \tag{2.19}$$

This simply describes the continuum limit of classical mechanics. The result of which is waves propagating in this system.

2.2 From quantum mechanics to Quantum Field Theory

In taking the continuum limit, we have found that

$$\delta y_j(t) \to \frac{1}{\sqrt{ka}} \phi(t,x) \,, \qquad \delta p_j(t) \to a\sqrt{ka}\,\pi(t,x) \,, \tag{2.20}$$

so that we can quantize the field ϕ starting from the canonical commutation relation

$$[\delta y_j(t), \delta p_{j'}(t)] = i\hbar \, \delta_{j,j'} \,, \tag{2.21}$$

obtaining

$$[\phi(t,x), \pi(t,x')] = \frac{1}{a} [\delta y_j(t), \delta p_{j'}(t)] = i\hbar \, \frac{\delta_{j,j'}}{a} \,. \tag{2.22}$$

The correct continuum identification turns the right-hand side into a Dirac delta function. In fact, from the continuum limit of the sum, $\sum_j \to \int \frac{dx}{a}$, we get

$$\sum_j \delta_{j,j'} = 1 \to \int \frac{dx}{a} \delta_{j,j'} = \int dx \, \delta(x - x') = 1. \qquad (2.23)$$

The end result is the commutator for field quantization

$$\boxed{[\phi(t, x), \pi(t, x')] = i\hbar \, \delta(x - x')} , \qquad (2.24)$$

which is the starting point of what is referred to as *canonical quantization* in Quantum Field Theory.

Like all things quantum, this rule takes some getting used to. It is saying that the field, which we can visualize classically as a wave propagating in front of us, is no longer just a function, but is an operator. There are a few things to say about this. We note that as we progressed in quantum mechanics we have become used to coordinates and momenta—also things that we have a good picture for classically—being operators as in equation (2.21). So at this moment, we will take a deep breath and just wait and see where this leads. In practice, it leads to a final result that is even easier to come to grips with than the usual quantum-mechanical formalism is. We will see that the field-operator formalism is really a bookkeeping device for keeping track of the creation and annihilation of particles.[2] That is actually much easier than the usual statement that position x is an operator. The coordinate x that appears in the quantum field–theoretical treatment of the system, for instance in equation (2.24), is *not* an operator, as it descends from the index j in the discrete system in equation (2.1). Finally, we should note that in the path integral formalism (see chapter 8), the fields again are treated as functions, and there is not an operator in sight. For now, you are counselled to be patient.[3]

2.3 Creation operators and the Hamiltonian

Now let us figure out how to solve the commutation rule in equation (2.24). To do this, we need to establish the general solutions to the wave equation. We propose doing this using the *box normalization*, in which the system is taken to be finite but very large, with length L. The exact boundary conditions are not important, but you can think of periodic boundary conditions for definiteness. What is useful about this choice is that energy levels are discrete, with a label n running on all the integers. This avoids having to simultaneously introduce the continuum momenta notation on top of the other ideas discussed in this section.

---

[2]For the reader who is not a native English speaker, a bookkeeper is one who keeps track of financial transactions. The use in the present context is that the field operators keep track of the physics transitions.

[3]We are also postponing to section 2.3 the reason that both equations (2.21) and (2.24) are evaluated at equal times even though the positions are different. Right now we are rushing to reach our goal—quanta.

The solutions of the wave equation, equation (2.12), take the form

$$\phi(t, x) = N\, e^{\pm i(\omega_n t - k_n x)} \quad \text{with} \quad \omega_n = |k_n| v, \qquad k_n = \frac{2\pi n}{L}, \tag{2.25}$$

where N is a arbitrary constant, so that the most general solution will be a superposition

$$\phi(t, x) = \sum_n N_n [\hat{a}_n\, e^{-i(\omega_n t - k_n x)} + \hat{a}_n^\dagger\, e^{+i(\omega_n t - k_n x)}], \tag{2.26}$$

where we have used the fact that ϕ is a real field. The coefficients $\hat{a}_n$ are now to be considered operators because ϕ is an operator.[4] The normalization factor N_n will be determined in equation (2.30).

Because $\hat{a}_n$ and $\hat{a}_n^\dagger$ are now operators, they must obey some commutation rules. If the different modes are to be orthogonal, we expect operators with different n values to commute. This leads to a set of rules that up to an overall normalization, reads

$$[\hat{a}_n, \hat{a}_{n'}] = 0, \tag{2.27}$$

$$[\hat{a}_n, \hat{a}_{n'}^\dagger] = \delta_{n,n'}. \tag{2.28}$$

The arbitrary overall normalization can be absorbed in the N_n factor in equation (2.26). This choice in fact does provide a solution to the field commutator rule, as we can readily see

$$[\phi(t, x), \pi(t, x')] = \sum_{n,n'} N_n N_{n'} \left(-i\frac{\omega_{n'}}{v^2}\right)$$

$$\times [\hat{a}_n e^{-i(\omega_n t - k_n x)} + \hat{a}_n^\dagger e^{+i(\omega_n t - k_n x)}, \; \hat{a}_{n'} e^{-i(\omega_{n'} t - k_{n'} x')} - \hat{a}_{n'}^\dagger e^{+i(\omega_{n'} t - k_{n'} x')}]$$

$$= \sum_{n,n'} N_n N_{n'} \left(2i\frac{\omega_{n'}}{v^2}\right) [\hat{a}_n, \hat{a}_{n'}^\dagger] e^{i(k_n x - k_{n'} x')}$$

$$= \sum_n \frac{i\hbar}{L} e^{ik_n(x-x')} = i\hbar\, \delta(x - x'), \tag{2.29}$$

provided that we take the normalization factor to be

$$N_n = \sqrt{\frac{\hbar v^2}{2\,\omega_n L}}. \tag{2.30}$$

The delta function identity

$$\delta(x - x') = \sum_n \frac{1}{L} e^{ik_n(x-x')} \tag{2.31}$$

follows from the completeness of the Fourier series.

---

[4]These coefficients are not to be confused with the length a between the mass points.

One could equivalently derive the commutation relation of the $\hat{a}_n$ and $\hat{a}_n^\dagger$ operators directly from the canonical quantization condition in equation (2.24) and invert the relations that give $\phi(t, x)$ and $\pi(t, x)$ as a function of $\hat{a}_n$ and $\hat{a}_n^\dagger$. More explicitly, we can rewrite equation (2.26) as

$$\phi(t, x) = \sum_n N_n \, e^{ik_n x} [\hat{a}_n e^{-i\omega_n t} + \hat{a}_{-n}^\dagger e^{+i\omega_n t}], \tag{2.32}$$

where we have used $\omega_{-n} = \omega_n$ and have assumed $N_n = N_{-n}$. Then, by inverting the Fourier series we obtain

$$\hat{a}_n = \frac{e^{i\omega_n t}}{2 N_n} \int \frac{dx}{L} \, e^{-ik_n x} \left[\phi(t, x) + iv^2 \frac{\pi(t, x)}{\omega_n} \right], \tag{2.33}$$

so that

$$
\begin{aligned}
[\hat{a}_n, \hat{a}_{n'}^\dagger] &= \frac{e^{i(\omega-\omega_{n'})t}}{4 N_n N_{n'}} \int \frac{dx \, dx'}{L^2} \, e^{-ik_n x + ik_{n'} x'} \\
&\quad \left[\phi(t, x) + iv^2 \frac{\pi(t, x)}{\omega_n}, \phi(t, x') - iv^2 \frac{\pi(t, x')}{\omega_{n'}} \right] \\
&= \frac{e^{i(\omega-\omega_{n'})t}}{4 N_n N_{n'}} \int \frac{dx \, dx'}{L^2} \, e^{-ik_n x + ik_{n'} x'} \, \hbar v^2 \left[\frac{\delta(x - x')}{\omega_{n'}} + \frac{\delta(x - x')}{\omega_n} \right] \\
&= \frac{\hbar v^2}{2 N_n^2 L \omega_n} \delta_{nn'} \tag{2.34}
\end{aligned}
$$

and, in similar fashion, $[\hat{a}_n, \hat{a}_{n'}] = [\hat{a}_n^\dagger, \hat{a}_{n'}^\dagger] = 0$. By choosing the normalization in equation (2.30), we find that each of the $\hat{a}_n, \hat{a}_n^\dagger$ pair of operators satisfy the same algebra as the creation and annihilation operators of the simple harmonic oscillator. Indeed we will find that this identification is accurate and will henceforth call $\hat{a}_n^\dagger$ a creation operator and $\hat{a}_n$ an annihilation operator.

We are now in position to evaluate the Hamiltonian. Because ϕ has two terms, $\hat{a}_n$ and $\hat{a}_n^\dagger$, the Hamiltonian

$$
\begin{aligned}
H &= \int dx \left[\frac{1}{2v^2} \left(\frac{\partial \phi}{\partial t} \right)^2 + \frac{1}{2} \left(\frac{\partial \phi}{\partial x} \right)^2 \right] \\
&= \int dx \sum_{n,n'} N_n N_{n'} \\
&\quad \times \left[\frac{-\omega_n \omega_{n'}}{2v^2} \left(\hat{a}_n e^{-i\psi_n} - \hat{a}_n^\dagger e^{+i\psi_n} \right) \left(\hat{a}_{n'} e^{-i\psi_{n'}} - \hat{a}_{n'}^\dagger e^{+i\psi_{n'}} \right) \right. \\
&\quad \left. - \frac{k_n k_{n'}}{2} \left(\hat{a}_n e^{-i\psi_n} - \hat{a}_n^\dagger e^{+i\psi_n} \right) \left(\hat{a}_{n'} e^{-i\psi_{n'}} - \hat{a}_n^\dagger e^{+i\psi_{n'}} \right) \right] \tag{2.35}
\end{aligned}
$$

will have four pieces, although they come in pairs because H is Hermitian. In the exponents, we have used the shorthand $\psi_n = (\omega_n t - k_n x)$ to save space.

The integral over x will imply that the momenta are either equal or opposite. By using

$$\int dx\, e^{ik_n x}\, e^{-ik_{n'} x} = L\, \delta_{n,n'}\,,$$

$$\int dx\, e^{ik_n x}\, e^{ik_{n'} x} = L\, \delta_{n,-n'}\,, \tag{2.36}$$

we can write equation (2.35) as a single sum over the momentum variable. Both of these cases have $\omega_n = \omega_{n'}$, and by inserting the normalization factor, we have

$$H = \sum_n \frac{\hbar v^2}{2\,\omega_n} \left[\frac{1}{2} \left(-\frac{\omega_n^2}{v^2} + k_n^2 \right) \left(e^{-2i\omega_n t} \hat{a}_n\, \hat{a}_{-n} + e^{+2i\omega_n t} \hat{a}_n^\dagger\, \hat{a}_{-n}^\dagger \right) \right.$$
$$\left. + \frac{1}{2} \left(+\frac{\omega_n^2}{v^2} + k_n^2 \right) \left(\hat{a}_n\, \hat{a}_n^\dagger + \hat{a}_n^\dagger\, \hat{a}_n \right) \right]. \tag{2.37}$$

At this stage "a miracle occurs" and the $\hat{a}_n\, \hat{a}_{-n}$ and $\hat{a}_n^\dagger \hat{a}_{-n}^\dagger$ terms disappear because $\omega_n^2 = k_n^2 v^2$. If we use the creation operator commutation rule, we obtain

$$H = H_0 + E_0\,, \tag{2.38}$$

with

$$\boxed{H_0 = \sum_n \hbar\,\omega_n\, \hat{a}_n^\dagger\, \hat{a}_n} \tag{2.39}$$

and

$$E_0 = \sum_n \frac{1}{2} \hbar\,\omega_n\,. \tag{2.40}$$

Here E_0 is the *zero-point energy*, which we will discuss in section 3.6. It provides a constant shift in energy that we will ignore for now. The other part of the Hamiltonian H_0 is quite promising. We see the *number operator* for each mode $\hat{a}_n^\dagger\, \hat{a}_n$ emerging, with an associated energy $E_n = \hbar\,\omega_n$. Note that it was the field commutation relation that fixed the normalization and, therefore, required the energy of the n-th mode to be $\hbar\,\omega_n$.

2.4 States filled with quanta

Once we identify the energy eigenstates of the theory and confirm that $\hat{a}_n^\dagger\, \hat{a}_n$ acts like a number operator, we will have reached our goal. This is easily fulfilled by using our experience with the simple harmonic oscillator. The states can be constructed by defining an "empty" state—the *vacuum*, which by definition is annihilated by all the annihilation operators

$$\hat{a}_n|0\rangle = 0 \quad \text{for all } n. \tag{2.41}$$

After this, we can construct new states by acting with the $\hat{a}_n$ and $\hat{a}_n^\dagger$ operators on the vacuum $|0\rangle$. Let us start with the action of a single $\hat{a}_n^\dagger$ operator, and let us define

$$|n\rangle = \hat{a}_n^\dagger|0\rangle. \tag{2.42}$$

This operation produces energy eigenstates, as can be readily verified

$$H|n\rangle = \sum_{n'} \hbar\omega_{n'} \hat{a}_{n'}^\dagger \hat{a}_{n'} (\hat{a}_n^\dagger|0\rangle) \tag{2.43}$$

$$= \sum_{n'} \hbar\omega_{n'} \hat{a}_{n'}^\dagger ([\hat{a}_{n'}, \hat{a}_n^\dagger] + \hat{a}_n^\dagger \hat{a}_{n'})|0\rangle \tag{2.44}$$

$$= \sum_{n'} \hbar\omega_{n'} \hat{a}_{n'}^\dagger \delta_{n,n'}|0\rangle = \hbar\omega_n|n\rangle. \tag{2.45}$$

This construction gives a state with energy $\hbar\omega_n$. We are thus led to interpret $|n\rangle$ as a *single particle state* that contains one quantum of the state with energy $\hbar\omega_n$. Here the word *particle* is used to mean a quantum carrying energy $\hbar\omega$. We will see later that we can define additional operators associated to observables such as momentum, charge, etc., and that the action of a creation operator on the vacuum gives eigenstates of all these operators, which is exactly as expected for a single particle state with those quantum numbers. For this reason, from now on, we will use interchangeably the terms "state $|n\rangle$" and "particle in n-th state."

On a technical point, we have explicitly worked out the action of the commutator when using the Hamiltonian, which is how the calculation proceeds. However, the way to think about this calculation is to mentally say "the annihilation operator $\hat{a}_{n'}$ annihilates the particle n'." This can be represented pictorially with a *contraction*

$$\sum_{n'} \hat{a}_{n'}^\dagger \underbrace{\hat{a}_{n'}|n\rangle} = \sum_{n'} \hat{a}_{n'}^\dagger \underbrace{\hat{a}_{n'}\hat{a}_n^\dagger}|0\rangle = \sum_{n'} \hat{a}_{n'}^\dagger [\hat{a}_{n'}, \hat{a}_n^\dagger]|0\rangle = \sum_{n'} \hat{a}_{n'}^\dagger \delta_{n,n'}|0\rangle = |n\rangle, \tag{2.46}$$

indicating that the given annihilation operator removes the creation operator. The calculation is given by the commutator but the result is indicated by the contraction. To acquire familiarity with contraction, you should work out the slightly more difficult case with two quanta

$$H|n_1, n_2\rangle = \sum_{n'} \hbar\omega_n' \hat{a}_{n'}^\dagger \underbrace{\hat{a}_{n'}|n_1, n_2\rangle} + \sum_{n'} \hbar\omega_n' \hat{a}_{n'}^\dagger \underbrace{\hat{a}_{n'}|n_1, n_2\rangle}$$

$$= (\hbar\omega_1 + \hbar\omega_2)|n_1, n_2\rangle. \tag{2.47}$$

of a single field. This explains why our Universe contains so many identical electrons, for instance: those electrons are not many different particles, but they are many different excitations of a single field. This is nearly the same as going from a description of the sea as a set of many waves to a description where it is a single body of water carrying a number of waves.

2.5 Connection with normal modes

The example we started with (the Lagrangian in equation (2.1)), was a specific system, but our analysis is actually valid for all many body systems near equilibrium. It is sometimes said that "a quantum field is an infinite number of harmonic oscillators." This can be seen from the expression for the Hamiltonian. It is also perhaps useful to go back to the discrete case and carry out the quantization procedure before taking the continuum limit. This is a solution via the normal mode technique.

If we start from the most general Lagrangian describing a system on N degrees of freedom near equilibrium,

$$L = \sum_i \left[\frac{m}{2} \dot{y}_i^2 - V(y_i) \right] \tag{2.50}$$

with a potential

$$V = \frac{1}{2} \sum_{ij} v_{ij}\, y_i\, y_j\,, \tag{2.51}$$

where v_{ij} is a real, symmetric $N \times N$ matrix, then this system can be solved by using normal mode techniques. The normal frequencies ω_n are the entries of the diagonal, positive, $N \times N$ matrix Ω found by solving

$$\det(m\,\Omega^2 - v) = 0\,. \tag{2.52}$$

The normal coordinates are then found via the modal matrix A_{in}

$$y_j = \sum_n A_{jn}\, \xi_n \quad \text{or} \quad \xi_n = \sum_j (A^T)_{nj}\, y_j\,. \tag{2.53}$$

This procedure decouples the harmonic oscillators

$$L = \sum_n \left[\frac{1}{2}\dot{\xi}_n^2 - \frac{1}{2}\omega_n^2\, \xi_n^2 \right]\,. \tag{2.54}$$

Each normal mode would then have an independent solution

$$\xi_n(t) = N\, (a_n e^{-i\omega_n t} + a_n^* e^{i\omega_n t}) \tag{2.55}$$

and the general solution would be a mixture of normal modes

$$y_j(t) = \sum_n A_{jn}\, \xi_n(t)\,. \tag{2.56}$$

Quantization then takes place independently for each normal mode

$$p_n = \frac{\partial L}{\partial \dot{\xi}_n} = \dot{\xi}_n \quad \text{with} \quad [\xi_n, p_{n'}] = i\hbar\, \delta_{n,n'}\,. \tag{2.57}$$

The coefficients in the normal mode expansion now need to become operators. Choosing

$$\xi_n = \sqrt{\frac{\hbar}{2\,\omega_n}}\, (\hat{a}_n\, e^{-i\omega_n t} + \hat{a}_n^\dagger\, e^{i\omega_n t})$$

and

$$p_n = -i\sqrt{\frac{\hbar\,\omega_n}{2}}\, (\hat{a}_n\, e^{-i\omega_n t} - \hat{a}_n^\dagger\, e^{i\omega_n t}) \tag{2.58}$$

and imposing the commutation rules $[\hat{a}_n, \hat{a}_{n'}^\dagger] = \delta_{n,n'}$, this leads to the Hamiltonian

$$H = \sum_n \hbar\omega_n \left(\hat{a}_n^\dagger \hat{a}_n + \frac{1}{2} \right) \tag{2.59}$$

and states as described in section 2.4.

The lesson of this exercise is that the states are the quanta of the normal modes. In field theory, the normal modes are wave solutions $A_{jn} \to e^{ik_n x}$ with the continuum identification $x = ja$. In the continuum limit the number of normal modes becomes infinite, hence the identification of the field quantization with an infinite number of normal modes.

Chapter summary: You have done it! You now understand how the usual rules of quantum mechanics lead to quanta of a field. We have found the commutation rules for fields and have seen how they can be expressed in terms of creation/annihilation operators and that can lead to an intuitive construction of the states of the system. We have defined the theory starting from the Lagrangian and, by following the rules, have expressed the related Hamiltonian in terms of the number operator. These are some of the most important lessons of Quantum Field Theory. There is much more to explore.

CHAPTER 3

Developing free field theory

There is much more to say! We rushed through the development of the field concept to quickly demonstrate that ordinary quantum mechanics inevitably leads to the concept of quanta for a field. Now we have to slow down and reflect on what we have done. The goal is to emerge from the present chapter feeling more comfortable about the idea of quantum fields.

3.1 Quantum mechanics in field theory notation

One of the oddities of the quantization procedure is that it has turned a field, which we think of as a simple function of space and time, into an operator. The fact that, while learning quantum mechanics, we have previously survived the equivalent transition of position and momenta into operators probably is not much comfort at the moment. However, in the end the operator nature of fields turns out to be easier to digest. We can view the operator part of the field as merely bookkeeping devices. This can be most easily seen by recasting usual quantum mechanics using these operators—a process referred to as *second quantization*.

Quantum physics often encodes the properties of a state $|j\rangle$ into a wave function

$$\langle x|j\rangle = \psi_j(x)\,, \qquad (3.1)$$

where the index j is a shorthand for all the quantum numbers that can define the state (in the example of chapter 1 this was just the wavenumber n). Let us assume that the state $|j\rangle$ is the eigenstate, with eigenvalue E_j, of some Hamiltonian. We can pack the complete set of these wavefunctions into a single field by defining the operator

$$\Psi(t, x) = \sum_j \psi_j(x)\, e^{-i\omega_j t}\, \hat{b}_j\,, \qquad \omega_j \equiv \frac{E_j}{\hbar}\,, \qquad (3.2)$$

where $\hat{b}_j$ is an annihilation operator that is defined by the usual commutation relations

$$[\hat{b}_j, \hat{b}_k^\dagger] = \delta_{jk} \quad \text{and} \quad [\hat{b}_j, \hat{b}_k] = [\hat{b}_j^\dagger, \hat{b}_k^\dagger] = 0 \qquad (3.3)$$

for bosons. For fermions we would use anticommutators, but let us not get into that at the moment.

The states for this theory can be defined using these creation operators, as in chapter 2,

$$|j\rangle = \hat{b}_j^\dagger \, |0\rangle, \quad \text{with} \quad \hat{b}_j \, |0\rangle = 0 \quad \forall j, \tag{3.4}$$

such that the field operating on the state returns the wavefunction

$$\Psi(t, x) \, |j\rangle = \psi_j(x) \, e^{-i\omega_j t} |j\rangle \,. \tag{3.5}$$

Even though we have constructed the field operator by including in it *all* the solutions, the field only retrieves the appropriate solution when acting on the corresponding state. In other words, we have moved the information on the state of the system from the wavefunction back to the state on which the field is acting.

In this notation, the Hamiltonian would be written as an integral over the fields

$$H = \int dx \, \Psi^\dagger(t, x) \left[-\frac{\hbar^2 \nabla^2}{2m} + V(x) \right] \Psi(t, x) \,, \tag{3.6}$$

as we can see by verifying that, because the wavefunctions $\psi_j(x)$ are orthonormal eigenfunctions of the Hamiltonian, we have

$$H = \sum_{j,k} \int dx \, \psi_j^*(x) \, \hat{b}_j^\dagger \, e^{i\omega_j t} [E_k \, e^{-i\omega_k t} \, \psi_k(x) \, \hat{b}_k] = \sum_j E_j \, \hat{b}_j^\dagger \, \hat{b}_j \,, \tag{3.7}$$

implying

$$H|j\rangle = E_j|j\rangle \,, \tag{3.8}$$

as expected.

Transitions between states can also be readily handled in this notation. Let us assume, for example, that there is an additional term in the potential, $V_I(x)$, that is not part of the original Hamiltonian for the energy eigenstates and is not diagonal in the original basis. We can still write it using the field operators,

$$H_I = \int dx \, \Psi^*(t, x) \, V_I(x) \, \Psi(t, x) \,, \tag{3.9}$$

with the result

$$H_I = \sum_{j,k} v_{jk} \, \hat{b}_j^\dagger \, \hat{b}_k \,, \quad \text{with} \quad v_{jk} \equiv \int dx \, \psi_j^*(x) \, V_I(x) \, \psi_k(x) e^{-i(\omega_k - \omega_j)t} \,, \tag{3.10}$$

corresponding to the *matrix element*

$$\langle j|H_I|k\rangle = v_{jk} \,. \tag{3.11}$$

Again, the field representation of the Hamiltonian contains all matrix elements and returns the appropriate one when acting between states. This is good bookkeeping.

Finally, we can anticipate the interaction section and look to see how we might include more complicated processes, such as atomic decay with photon emission. The interaction with a photon associated to a vector potential $\mathbf{A}(t, \mathbf{x})$ is given by $H_I = -\frac{e}{m}\mathbf{p} \cdot \mathbf{A}$, which in this notation would be represented by

$$H_I = \int d^3x \left[\Psi^*(t, \mathbf{x}) \frac{ie\,\nabla}{m} \Psi(t, \mathbf{x}) \right] \cdot \mathbf{A}(t, \mathbf{x}). \qquad (3.12)$$

Although we have not yet discussed how we would treat the photon field, we will eventually describe its states by using creation operators. In the matrix element

$$\langle j, \gamma(\mathbf{p})|H_I|k \rangle \qquad (3.13)$$

the operators would do the bookkeeping and return to us an amplitude for the transition describing the decay of a mode in the state $|k\rangle$ into a mode in the state $|j\rangle$ and a photon of momentum $\mathbf{p}$.

In quantum mechanics, the only important things are the amplitudes. The operator nature of the field disappears in the end and leaves behind the appropriate matrix element. Along the way, it is a handy method to keep track of the states of the theory and to organize the possible transitions among those states.

3.2 The infinite-box limit

So far we have considered a field emerging from the continuum description of a string of finite length. This, in particular, has implied that the mode decomposition of the field was performed in terms of a Fourier series. However, it is more realistic, and in most cases simpler, to assume that the system lives in a box of infinite size, which will lead to the use of Fourier transforms instead of Fourier series.

The infinite-box limit is obtained in a straightforward way as follows. We start from the three-dimensional generalization of equations (2.26) and (2.30),

$$\phi(t, \mathbf{x}) = \sum_{\mathbf{n}} \sqrt{\frac{\hbar v^2}{2\omega_{|\mathbf{n}|} L^3}} \left[\hat{a}_{\mathbf{n}}\, e^{-i(\omega_{|\mathbf{n}|}t - 2\pi \mathbf{n} \cdot \mathbf{x}/L)} + \hat{a}_{\mathbf{n}}^\dagger\, e^{+i(\omega_{|\mathbf{n}|}t - 2\pi \mathbf{n} \cdot \mathbf{x}/L)} \right], \qquad (3.14)$$

then we make the replacement $\sum_{\mathbf{n}} \to \int d^3n = L^3 \int \frac{d^3k}{(2\pi)^3}$, where we have defined $\mathbf{k} = 2\pi\mathbf{n}/L$. We then redefine $L^{3/2}\hat{a}_{\mathbf{n}} \to \hat{a}_{\mathbf{k}}$. This brings us to the expression

$$\phi(t, \mathbf{x}) = \int \frac{d^3k}{(2\pi)^3} \sqrt{\frac{\hbar v^2}{2\omega_k}} \left[\hat{a}_{\mathbf{k}}\, e^{-i(\omega_k t - \mathbf{k} \cdot \mathbf{x})} + \hat{a}_{\mathbf{k}}^\dagger\, e^{+i(\omega_k t - \mathbf{k} \cdot \mathbf{x})} \right], \qquad (3.15)$$

where the new creation/annihilation operators obey the algebra

$$[\hat{a}_{\mathbf{k}}, \hat{a}_{\mathbf{k}'}^{\dagger}] = \frac{1}{L^3}\delta^{(3)}\left(\frac{L}{2\pi}\mathbf{k} - \frac{L}{2\pi}\mathbf{k}'\right) = (2\pi)^3\,\delta^{(3)}(\mathbf{k} - \mathbf{k}')\,. \tag{3.16}$$

We will thus denote the state with a single particle of momentum $\mathbf{p}$ as $|\mathbf{p}\rangle = \hat{a}_{\mathbf{p}}^{\dagger}|0\rangle$, and it is easy to check that commutation relation (3.16) implies the normalization $\langle\mathbf{p}'|\mathbf{p}\rangle = (2\pi)^3\,\delta^{(3)}(\mathbf{p}' - \mathbf{p})$.

The normalization of the states is an area where there are different conventions in the literature. Our choice

$$\langle\mathbf{p}'|\mathbf{p}\rangle = (2\pi)^3\delta^{(3)}(\mathbf{p}' - \mathbf{p}) \tag{3.17}$$

has the advantage that

$$\langle\mathbf{p}'|H|\mathbf{p}\rangle = E_p\,(2\pi)^3\delta^{(3)}(\mathbf{p}' - \mathbf{p})\,. \tag{3.18}$$

However, it leaves various $1/\sqrt{2E}$ factors floating around in matrix elements (i.e., note the factor of $1/\sqrt{2\omega}$ in equation (3.15)). We will remove these when we define the Feynman rules in chapter 5. When they are stripped off in this fashion, they reemerge in the derivation of the formulas for the decay width and the cross section in a way that is natural.

The other common choice of normalization

$$\langle\mathbf{p}'|\mathbf{p}\rangle = 2E_p\,(2\pi)^3\,\delta^{(3)}(\mathbf{p}' - \mathbf{p}) \tag{3.19}$$

would get rid of the $1/\sqrt{2E}$ factors earlier. This makes the identification of the Feynman rules simpler, as these factors do not need to be discussed. It makes the derivation of the decay width and cross section a bit more complicated, because it is less obvious how these factors appear in those derivations. It also makes the early discussion of the energy slightly odd because we have

$$\langle\mathbf{p}'|H|\mathbf{p}\rangle = 2\,E_p^2\,(2\pi)^3\,\delta^{(3)}(\mathbf{p}' - \mathbf{p})\,. \tag{3.20}$$

However, it is a very common choice of normalization for states, and you will likely encounter it in some settings.

3.3 Relativistic notation, $\hbar = c = 1$, and dimensional analysis

The factors $\hbar$ and v that appear throughout the formulas add to the clutter of the notation. To see what is essential, it is useful to remove inessential symbols and to simplify the notation as much as reasonable. For relativistic field theory, the velocity v equals the speed of light, c. For most of the rest of the book we will use notation where $\hbar$ and c are equal to unity. We will also use relativistic notation whenever possible (if you are reading this book, you have probably seen enough of Special

Relativity, so we will not review this here). Quantum Field Theory is also important in the nonrelativistic realm. But relativistic notation is more compact and is important in its own context. This section is designed to get you comfortable with these changes.

After having set the speed of light to unity, $c = 1$, an event happening at time t and location $\mathbf{x}$ will be denoted by the four-vector $x^\mu \equiv (t, \mathbf{x})$. For fields that are functions of these variables, we will often not display the time and space variables separately, instead we will use $\phi(x) = \phi(t, \mathbf{x})$. Lorentz invariant objects are formed with an extra minus sign, and we use the "mostly minus" convention written variously as

$$x^2 = t^2 - \mathbf{x}^2 = x^\mu x_\mu = x \cdot x = x^\mu \, \eta_{\mu\nu} \, x^\nu \,. \tag{3.21}$$

The last form introduces the Minkowski metric tensor

$$\eta_{\mu\nu} = \mathrm{diag}(1, \, -1, \, -1, \, -1) \,. \tag{3.22}$$

The four-gradient will then take the form $\partial_\mu \equiv (\partial_t, \, \boldsymbol{\nabla})$. Our choice, equation (3.22), then implies that $\partial^\mu = (\partial_t, \, -\boldsymbol{\nabla})$, leading to the D'Alembertian

$$\Box \equiv \partial_\mu \partial^\mu = \partial_t^2 - \boldsymbol{\nabla}^2 \,, \tag{3.23}$$

whereas the kinetic energy terms in the Lagrangian density of the scalar field (the equivalent of equation (2.9)) takes the symmetric form

$$\mathcal{L} = \frac{1}{2} \, \partial_\mu \phi \, \partial^\mu \phi + \dots \,. \tag{3.24}$$

We also define the four-momentum $p^\mu = (E, \mathbf{p})$ that satisfies, for a particle of mass m, the relation

$$p^\mu p_\mu = p \cdot p = E^2 - \mathbf{p}^2 = m^2 \,, \tag{3.25}$$

and the phase appearing in relativistic plane waves takes the compact form $E t - \mathbf{p} \cdot \mathbf{x} = p^\mu x_\mu = p \cdot x$. Finally, we will use the fact that the spacetime volume element $dt \, d^3 x = d^4 x$ is a relativistic invariant. Under a boost with velocity $\mathbf{v}$, time is dilated by a factor $\gamma \equiv 1/\sqrt{1 - \mathbf{v}^2}$ and one space dimension is contracted by a factor γ^{-1}, with the remaining two dimensions unchanged.

Note that the mostly minus convention, equation (3.22), we use here is the standard one in particle physics, as opposed to the "mostly plus" convention that is typical of the general theory of relativity. The reason is that, in mostly minus convention, equation (3.25) reads $p^\mu p_\mu = +m^2$. This is an important relation in particle physics, and the comfort of not having to deal with a minus sign on the right-hand side of that equation compensates the uneasiness we can feel by having the minus sign in front of the spacelike coordinates.

Because the speed of light c and the Planck constant $\hbar$ are *universal* constants, we can set them equal to unity. This implies that the three fundamental units (of length, of time, and of mass) reduce to a single unit. For instance, because

$c = 1 = 299{,}792{,}458$ m/s, we can write $1\text{s} = 299{,}792{,}458$ m, so that length and time have the same units. From the fundamental commutation relation $[x, p] = i\hbar$, we also discover that momentum has the dimensions of inverse length. And from $E^2 = m^2 c^4 + p^2 c^2$, we find that both mass and momenta have dimensions of energy. It is customary to use the unit of energy $[E]$ as the only fundamental unit, so that

$$[\text{momentum}] = [\text{energy}] = [\text{mass}] = [E], \qquad [\text{length}] = [\text{time}] = [E^{-1}]. \quad (3.26)$$

A useful conversion factor is $\hbar c = 197$ eV nm $= 197$ MeV fm. Because this factor is unity in natural units, it can be inserted wherever it is useful to convert energy units in distance units. That factor plus the value of c are sufficient to recover physical units from any expression.

3.4 Action principle in general

By generalizing the example analyzed in chapter 2, it is natural to conclude that the general action of a field will be the integral over spacetime of a Lagrangian density, which depends on the field and on its spacetime derivatives. Employing the relativistic notation that we have just introduced, we can write

$$S = \int d^4x \, \mathcal{L}(\partial_\mu \phi, \, \phi), \quad (3.27)$$

and the equations of motion will be obtained by imposing the condition that the action is stationary under small variations of the field about their solution $\bar{\phi}(x)$

$$
\begin{aligned}
0 = \delta S &\equiv \int d^4x \, \mathcal{L}\left(\partial_\mu\left(\bar{\phi} + \delta\phi\right), \, \bar{\phi} + \delta\phi\right) - \int d^4x \, \mathcal{L}\left(\partial_\mu\bar{\phi}, \, \bar{\phi}\right) \\
&= \int d^4x \left[\frac{\partial \mathcal{L}\left(\partial_\mu\bar{\phi}, \, \bar{\phi}\right)}{\partial(\partial_\mu\bar{\phi})} \partial_\mu \delta\phi + \frac{\partial \mathcal{L}\left(\partial_\mu\bar{\phi}, \, \bar{\phi}\right)}{\partial\bar{\phi}} \delta\phi \right] \\
&= \int d^4x \left[-\partial_\mu \frac{\partial \mathcal{L}\left(\partial_\mu\bar{\phi}, \, \bar{\phi}\right)}{\partial(\partial_\mu\bar{\phi})} + \frac{\partial \mathcal{L}\left(\partial_\mu\bar{\phi}, \, \bar{\phi}\right)}{\partial\bar{\phi}} \right] \delta\phi,
\end{aligned}
\quad (3.28)
$$

where in the last step we have performed an integration by parts, neglecting the boundary terms as we assume that $\delta\phi$ vanishes at both at spacelike and at timelike infinity.

Because the variation of the action in equation (3.28) must vanish for all functions $\delta\phi$, we get the *Euler-Lagrange equation*

$$\boxed{\; \partial_\mu \frac{\partial \mathcal{L}}{\partial(\partial_\mu \phi)} - \frac{\partial \mathcal{L}}{\partial\phi} = 0 \;} \;. \quad (3.29)$$

Let us make a couple of comments on this result.

First, equation (3.29) is Lorentz invariant if $\mathcal{L}$ is a Lorentz scalar. (This implies that, because d^4x is a Lorentz-invariant quantity, the action S is a Lorentz scalar.) This criterion will provide a great deal of guidance when we construct general Lagrangian densities in chapter 4. In a nonrelativistic field theory we will not expect full Lorentz invariance, but the Lagrangian should still be invariant under translations and rotations.

Second, while thus far we have assumed that the Lagrangian density depends only on ϕ and $\partial_\mu \phi$, it is straightforward to extend the derivation of equation (3.29) to Lagrangians with higher order derivatives. For instance, if $\mathcal{L} = \mathcal{L}(\phi, \partial_\mu \phi, \Box\phi)$, then the equation of motion will read

$$\Box \frac{\partial \mathcal{L}}{\partial(\Box\phi)} - \partial_\mu \frac{\partial \mathcal{L}}{\partial(\partial_\mu \phi)} + \frac{\partial \mathcal{L}}{\partial \phi} = 0 \,. \tag{3.30}$$

However, in this book we will deal with equations of motion, which are at most second order in the derivatives.

3.5 Energy and momentum

We have seen in chapter 2 that we can construct the Hamiltonian of a field starting from its Lagrangian density $\mathcal{L}(\phi, \dot{\phi}, \nabla\phi)$ as

$$H = \int d^3x \left[\left(\frac{\partial \mathcal{L}}{\partial \dot{\phi}} \right) \dot{\phi} - \mathcal{L} \right]. \tag{3.31}$$

From the study of classical mechanics we know that energy is conserved in systems with a finite number of degrees of freedom if the Lagrangian does not depend explicitly on time. How does this result carry over to fields? Let us see what happens when we compute the time derivative of the Hamiltonian in equation (3.31)

$$\frac{dH}{dt} = \int d^3x \left[\left(\frac{d}{dt}\frac{\partial \mathcal{L}}{\partial \dot{\phi}} \right) \dot{\phi} + \left(\cancel{\frac{\partial \mathcal{L}}{\partial \dot{\phi}}} \right) \ddot{\phi} - \left(\frac{\partial \mathcal{L}}{\partial \phi} \right) \dot{\phi} - \left(\cancel{\frac{\partial \mathcal{L}}{\partial \dot{\phi}}} \right) \ddot{\phi} - \left(\frac{\partial \mathcal{L}}{\partial(\nabla\phi)} \right) \cdot \nabla\dot{\phi} \right]$$

$$= \int d^3x \left[\left(\cancel{\frac{d}{dt}\frac{\partial \mathcal{L}}{\partial \dot{\phi}}} \right) \dot{\phi} - \left(\cancel{\frac{\partial \mathcal{L}}{\partial \phi}} \right) \dot{\phi} + \left(\nabla \cdot \cancel{\frac{\partial \mathcal{L}}{\partial(\nabla\phi)}} \right) \dot{\phi} - \nabla \cdot \left(\frac{\partial \mathcal{L}}{\partial(\nabla\phi)} \dot{\phi} \right) \right], \tag{3.32}$$

where we have performed an integration by parts when going from the first to the second line. The three terms in the second line cancel out by virtue of the equations of motion. Using Stokes's theorem we can then write the time derivative of the Hamiltonian as a boundary term

$$\frac{dH}{dt} = - \int d^2x \,\hat{\mathbf{n}} \cdot \left(\frac{\partial \mathcal{L}}{\partial(\nabla\phi)} \dot{\phi} \right) \tag{3.33}$$

and, assuming that the fields vanish at spatial infinity, we recover the conservation of the total energy in the system, $dH/dt = 0$.

While this derivation allowed us to show that the *total* Hamiltonian is constant, these same steps allow us to do better and find a *local* conservation law. To do this, let us remind ourselves that the total Hamiltonian of the system can be written as an integral over space of a Hamiltonian density $\mathcal{H}$, with $H = \int d^3x \, \mathcal{H}$. Then, because equation (3.32) is valid for any choice of volume over which it is integrated, it must hold also for the integrand, so that

$$\frac{d\mathcal{H}}{dt} + \mathbf{\nabla} \cdot \left(\frac{\partial \mathcal{L}}{\partial (\mathbf{\nabla}\phi)} \dot{\phi} \right) = 0. \tag{3.34}$$

But, what is $\frac{\partial \mathcal{L}}{\partial (\mathbf{\nabla}\phi)} \dot{\phi}$? To find this, let us notice that our derivation assumed that the Lagrangian does not depend *explicitly* on space and time. Then, let us consider the total derivative of the Lagrangian with respect to one coordinate x^α. Using the chain rule, we have

$$
\begin{aligned}
\frac{d\mathcal{L}}{dx^\alpha} &= \frac{\partial \mathcal{L}}{\partial \phi} \frac{\partial \phi}{\partial x^\alpha} + \frac{\partial \mathcal{L}}{\partial (\partial_\mu \phi)} \frac{\partial (\partial_\mu \phi)}{\partial x^\alpha} \\
&= \frac{\partial \mathcal{L}}{\partial \phi} \frac{\partial \phi}{\partial x^\alpha} + \partial_\mu \left(\frac{\partial \mathcal{L}}{\partial (\partial_\mu \phi)} \frac{\partial \phi}{\partial x^\alpha} \right) - \left(\partial_\mu \frac{\partial \mathcal{L}}{\partial (\partial_\mu \phi)} \right) \frac{\partial \phi}{\partial x^\alpha},
\end{aligned} \tag{3.35}
$$

where we have used the Euler-Lagrange equations to cancel the first and last term on the second line. We can rewrite the equation above as

$$\partial_\mu \left(\frac{\partial \mathcal{L}}{\partial (\partial_\mu \phi)} \partial_\alpha \phi - \delta^\mu_\alpha \mathcal{L} \right) = 0. \tag{3.36}$$

We thus discover that, if $\mathcal{L}$ does not depend on any of the coordinates x^α, we have four conserved currents, whose components correspond to the components of the energy-momentum tensor (also referred to as the stress-energy tensor)

$$\boxed{ T_{\mu\nu} = \frac{\partial \mathcal{L}}{\partial (\partial^\mu \phi)} \partial_\nu \phi - \eta_{\mu\nu} \mathcal{L} } . \tag{3.37}$$

In particular, we recover the energy density $\mathcal{H} = T_{00}$, and we see that the object appearing in equation (3.34) is

$$\frac{\partial \mathcal{L}}{\partial (\nabla_i \phi)} \dot{\phi} = T^i{}_0 = \mathcal{P}^i. \tag{3.38}$$

$\mathcal{P}^i$, being associated with invariance under space translations, is interpreted as *momentum density*.

We thus find that equation (3.34) describes the change in energy density as the flow of momentum: $\dot{\mathcal{H}} + \nabla \cdot \mathcal{P} = 0$.

Now, we have seen in chapter 2 that $\hat{a}_{\mathbf{p}}^{\dagger}$ creates a state $|\mathbf{p}\rangle$ that is an eigenstate of the Hamiltonian, $H|\mathbf{p}\rangle = E_p |\mathbf{p}\rangle$. What does the momentum operator do to the state $|\mathbf{p}\rangle$? By inserting the decomposition in equation (3.15) into the definition of momentum (but now with $v = c = 1$ and $\hbar = 1$!), we obtain, after some calculation,

$$\int d^3x \, \mathcal{P} = \int \frac{d^3q}{(2\pi)^3} \, \mathbf{q} \, \hat{a}_{\mathbf{q}}^{\dagger} \hat{a}_{\mathbf{q}} , \qquad (3.39)$$

so that

$$\int d^3x \, \mathcal{P} |\mathbf{p}\rangle = \int \frac{d^3q}{(2\pi)^3} \mathbf{q} \, \hat{a}_{\mathbf{q}}^{\dagger} \hat{a}_{\mathbf{q}} \, (\hat{a}_{\mathbf{p}}^{\dagger}|0\rangle) = \mathbf{p} |\mathbf{p}\rangle . \qquad (3.40)$$

Thus $|\mathbf{p}\rangle$ is a state of momentum $\mathbf{p}$, and, as we have seen, energy E_p. This should be enough to convince ourselves that it really describes a particle.

3.6 Zero-point energy

By repeating, in the infinite-box limit, the derivation of the Hamiltonian from section 2.3, we obtain

$$H = \int d^3x \left[\left(\frac{\partial \mathcal{L}}{\partial \dot{\phi}} \right) \dot{\phi} - \mathcal{L} \right] = \int \frac{d^3q}{(2\pi)^3} \frac{E_q}{2} (\hat{a}_{\mathbf{q}}^{\dagger} \hat{a}_{\mathbf{q}} + \hat{a}_{\mathbf{q}} \hat{a}_{\mathbf{q}}^{\dagger}) , \qquad (3.41)$$

where $E_q = \sqrt{\mathbf{q}^2 + m^2}$. By using the commutation relations in equation (3.16) we can rewrite the above equation as

$$H = \int \frac{d^3q}{(2\pi)^3} E_q \left(\hat{a}_{\mathbf{q}}^{\dagger} \hat{a}_{\mathbf{q}} + \frac{(2\pi)^3}{2} \delta_{\mathbf{q}}^{(3)}(0) \right) , \qquad (3.42)$$

where the symbol $\delta_{\mathbf{q}}^{(3)}(0)$ refers to a Dirac delta function in momentum space computed at zero argument.

Let us now focus on the divergent additive term in the above equation

$$E_0 \equiv \frac{1}{2} \int d^3q \, E_q \, \delta_{\mathbf{q}}^{(3)}(0) . \qquad (3.43)$$

We can make sense of the $\delta_{\mathbf{q}}^{(3)}(0)$ by writing it as a Fourier transform,

$$\delta_{\mathbf{q}}^{(3)}(0) = \int \frac{d^3x}{(2\pi)^3} \, e^{i\mathbf{q}\cdot\mathbf{x}} \Big|_{\mathbf{q}=0} = \frac{V}{(2\pi)^3} , \qquad (3.44)$$

where V is the (divergent) volume of the box containing the system. Also, the integral in d^3q in equation (3.43) is divergent in the ultraviolet, but if we cut it off at some high momentum Λ, assuming $\Lambda \gg m$, we have

$$E_0 \simeq \frac{1}{16\,\pi^2}\, V \Lambda^4 . \tag{3.45}$$

Where does this *zero-point energy* come from? Let us go back to the discrete system of chapter 2. Our initial system had N harmonic oscillators on a length $L = N\,a$. While the energy of the ground state of a classical harmonic oscillator vanishes, each quantum oscillator has an energy $\frac{1}{2}\sqrt{k/m}$ in its ground state. So the total energy of the ground state of the system is

$$E_0 = \frac{1}{2}\sqrt{\frac{k}{m}}\,N = \frac{1}{2}\sqrt{\frac{k}{m}\frac{L}{a}} = \frac{v}{2\,a^2}L , \tag{3.46}$$

where v is the speed of sound in the system (i.e., $c = 1$ in our case). The continuum limit of the one-dimensional version of equation (3.45) gives

$$E_0 \simeq \frac{L \Lambda^2}{8\,\pi} . \tag{3.47}$$

We thus see that equations (3.46) and (3.47) agree, up to an order unity factor, once we identify the ultraviolet cutoff Λ with $2\pi/a$, where a is the interparticle distance. In this book, in which we ignore the effects of gravity, we can (and will) ignore E_0 because adding a constant to the Hamiltonian does not affect any physics.

3.7 Noether's theorem

The conservation of energy and momentum in a theory where the Lagrangian does not depend explicitly on space and time, as seen in section 3.5, is a special case of the more general statement: if the action of a system is invariant under a set of continuous transformations, then the system contains a corresponding number of conserved quantities. This is *Noether's theorem*, which holds both for the transformations acting on spacetime coordinates of section 3.5 and for so-called *internal transformations* that involve the fields themselves. We will discuss this latter case now.

But we should not rush ahead and prove Noether's theorem for internal transformations without first stressing the importance of this result. We could see its importance just from the fact that it relates something as simple as invariance under time translations to the conservation of energy, a fact that has such an impact on the functioning of our Universe. More generally, because it relates conservation of certain quantities (a fact that can be observed experimentally) to symmetries in the action of the theory, Noether's theorem provides a powerful bridge from observations to theory. Given that symmetries have turned out, during the last century, to provide the main criterion in our construction of theories of elementary particles, the effects of Noether's theorem are ubiquitous.

Let us now prove Noether's theorem for an arbitrary number K of internal symmetries involving an arbitrary number N of fields. Our hypothesis is that the action S is invariant under the transformations

$$\phi_i \mapsto \phi_i + \sum_{\ell=1}^{K} F_{i\ell}(\phi, \partial_\mu \phi)\, \alpha_\ell \equiv \phi_i + \delta\phi_i \,, \qquad (3.48)$$

where α_ℓ are arbitrary but infinitesimal constants, and the $F_{i\ell}$ are functions of the fields and their derivatives. Invariance under the transformation in equation (3.48) implies that

$$0 = \delta S = \int d^4 x \sum_{i=1}^{N} \left(\frac{\partial \mathcal{L}}{\partial \phi_i} \delta\phi_i + \frac{\partial \mathcal{L}}{\partial(\partial_\mu \phi_i)} \partial_\mu \delta\phi_i \right)$$

$$= \int d^4 x \sum_{i=1}^{N} \sum_{\ell=1}^{K} \left(\frac{\partial \mathcal{L}}{\partial \phi_i} F_{i\ell} + \frac{\partial \mathcal{L}}{\partial(\partial_\mu \phi_i)} \partial_\mu F_{i\ell} \right) \alpha_\ell \qquad (3.49)$$

and, because the α_ℓ are arbitrary,

$$\int d^4 x \sum_{i=1}^{N} \left(\frac{\partial \mathcal{L}}{\partial \phi_i} F_{i\ell} + \frac{\partial \mathcal{L}}{\partial(\partial_\mu \phi_i)} \partial_\mu F_{i\ell} \right) = 0 \,, \quad \text{and} \quad \forall \ell = 1, \dots, K. \qquad (3.50)$$

Let us now consider a transformation of the action in the case in which the quantities α_ℓ are now infinitesimally small arbitrary functions of space and time: $\alpha_\ell = \alpha_\ell(x)$. This implies that the $\delta\phi_i$ in equation (3.48) are also arbitrary functions, and the operation in equation (3.48) now corresponds to a small variation of the fields that, courtesy the principle of least action, leaves the action invariant when the ϕ_is are solutions to the equations of motion:

$$0 = \delta S = \int d^4 x \sum_{i=1}^{N} \sum_{\ell=1}^{K} \left(\frac{\partial \mathcal{L}}{\partial \phi_i} F_{i\ell}\, \alpha_\ell(x) + \frac{\partial \mathcal{L}}{\partial(\partial_\mu \phi_i)} \partial_\mu (F_{i\ell}\, \alpha_\ell(x)) \right)$$

$$= \int d^4 x \sum_{i=1}^{N} \sum_{\ell=1}^{K} \left(\frac{\partial \mathcal{L}}{\partial(\partial_\mu \phi_i)} F_{i\ell}\, \partial_\mu \alpha_\ell(x) \right) \,, \qquad (3.51)$$

where in going from the first to the second line we have used equation (3.50). Integrating by parts the last line of the above equation, and using the fact that the functions $\alpha_\ell(x)$ are arbitrary, we finally obtain the conservation law

$$\boxed{\; \partial_\mu j_\ell^\mu = 0 \,, \quad \text{with} \quad j_\ell^\mu = \sum_{i=1}^{N} \frac{\partial \mathcal{L}}{\partial(\partial_\mu \phi_i)} F_{i\ell} \;} \,, \qquad (3.52)$$

which implies that the quantities

$$Q_\ell \equiv \int d^3x\, j_\ell^0, \quad \text{and} \quad \ell = 1, \dots, K \tag{3.53}$$

are constant, because

$$\frac{dQ_\ell}{dt} = \int d^3x\, \frac{dj_\ell^0}{dt} = -\int d^3x\, \boldsymbol{\nabla} \cdot \boldsymbol{j}_\ell = -\int_\Sigma d^2x\, \hat{\mathbf{n}} \cdot \boldsymbol{j}_\ell = 0. \tag{3.54}$$

In the second equality we have used equation (3.52); in the third equality we have used Stokes's theorem (where Σ is a two-dimensional surface that encloses all of space and $\hat{\mathbf{n}}$ is the vector normal to that surface); and in the last equality we have assumed that all fields and currents vanish at spatial infinity.

This concludes the proof of Noether's theorem: not only have we shown the existence of the conserved currents and the associated conserved charges, but equation (3.52) gives the explicit form of those currents. We will use this theorem in section 3.9.

3.8 The relativistic real scalar field

By establishing that the Lagrangian is be a Lorentz scalar and by requiring second-order differential equations, we find that the general Lagrangian for a real scalar field reads

$$\mathcal{L} = \frac{1}{2}\partial_\mu\phi\,\partial^\mu\phi - V(\phi), \tag{3.55}$$

where the use of $\frac{1}{2}$ in front of the kinetic term is conventional (it can be changed by multiplying ϕ by an arbitrary constant). The alert reader will look past the notation and recognize that the kinetic energy portion of this Lagrangian is similar to that of the one-dimensional phonon example that we described in chapter 2. The *potential* $V(\phi)$ describes interactions and will be discussed in more detail in chapter 4. We will generally focus on a potential of the form

$$V(\phi) = \frac{m^2}{2}\phi^2 + \frac{\lambda_3}{3!}\phi^3 + \frac{\lambda_4}{4!}\phi^4, \tag{3.56}$$

with m^2, λ_3, and λ_4 arbitrary real constants (later we will see that stability of the vacuum requires $\lambda_4 \geq 0$).

Note that $V(\phi)$ in general does not contain a constant term because adding a constant to the Lagrangian does not affect the equations of motion and does not contain a term linear in ϕ because such a term can be eliminated by redefining $\phi \mapsto \phi + \text{constant}$. We also neglect terms of the form $\lambda_n\phi^n$ with $n \geq 5$, whose coefficients have the dimensions $[\lambda_n] = [E]^{-(n-4)}$. We will return to such terms when we discuss renormalizability.

The equations of motion derived from the Lagrangian in equation (3.55) with the potential in equation (3.56) read

$$(\Box + m^2)\,\phi = -\frac{\lambda_3}{2}\phi^2 - \frac{\lambda_4}{3!}\phi^3 , \tag{3.57}$$

where the right-hand side is associated to the interactions in the theory. If these terms were to be neglected, the left-hand side would describe the free field equation of motion. Starting in chapter 4 we will focus on the effects of the interactions. Here we consider only the free field. Then, following the same procedures seen in chapter 2, we can quantize the field as

$$\phi(x) = \phi(t,\mathbf{x}) = \int \frac{d^3k}{(2\pi)^3}\frac{1}{\sqrt{2\omega_k}}\,[\hat{a}_\mathbf{k}\,e^{-i(\omega_k t - \mathbf{k}\cdot\mathbf{x})} + \hat{a}_\mathbf{k}^\dagger\,e^{+i(\omega_k t - \mathbf{k}\cdot\mathbf{x})}]$$

$$\omega_k \equiv \sqrt{k^2 + m^2} , \tag{3.58}$$

where $\hat{a}_\mathbf{k}$ and $\hat{a}_\mathbf{k}^\dagger$ satisfy the creation/annihilation operator commutation relations in equation (3.16) and where the energy operator is given by equation (3.42).

3.9 The complex scalar field and antiparticles

So far we have focused on the case of a single real scalar field. We can also consider systems containing more degrees of freedom. Let us move on to the next case in line: two real scalar fields, $\phi_1(x)$ and $\phi_2(x)$. Generalizing the discussion in the previous sections to the case of two fields is straightforward. Additionally, by increasing the number of degrees of freedom, we increase our possibilities of introducing more symmetries and, thanks to Noether's theorem, more conserved quantities.

In the case of two real fields of equal mass, it can be convenient to organize them into a *complex scalar field*

$$\phi(x) = \frac{1}{\sqrt{2}}\,(\phi_1(x) + i\,\phi_2(x)) . \tag{3.59}$$

The Euler-Lagrange equations will read

$$\partial_\mu \frac{\partial\mathcal{L}}{\partial(\partial_\mu\phi)} - \frac{\partial\mathcal{L}}{\partial\phi} = \partial_\mu \frac{\partial\mathcal{L}}{\partial(\partial_\mu\phi^*)} - \frac{\partial\mathcal{L}}{\partial\phi^*} = 0 . \tag{3.60}$$

The most general Lagrangian will now be a function of $\partial_\mu\phi$, $\partial_\mu\phi^*$, ϕ, and ϕ^*. A new symmetry, however, emerges if we stipulate that the Lagrangian depends only on the combinations $\partial_\mu\phi^*\partial^\mu\phi$ and $\phi^*\phi$. In this case the Lagrangian is invariant under the transformation $\phi \mapsto e^{i\alpha}\,\phi$ for a arbitrary real α. We will consider this to be the case from now on, so that the Lagrangian will read

$$\mathcal{L} = \partial_\mu\phi^*\,\partial^\mu\phi - m^2\,\phi^*\phi - \lambda_4(\phi^*\phi)^2 , \tag{3.61}$$

leading to the equation of motion

$$(\Box + m^2)\,\phi = -2\lambda_4\phi\,(\phi^*\phi)\,. \tag{3.62}$$

We can obtain the equation of the free complex scalar by neglecting the right-hand side of the above equation. We can thus quantize the field decomposing it as

$$\phi(x) = \int \frac{d^3k}{(2\pi)^3}\frac{1}{\sqrt{2\omega_k}}\,[\hat{a}_{\mathbf{k}}\,e^{-i(\omega_k t - \mathbf{k}\cdot\mathbf{x})} + \hat{b}_{\mathbf{k}}^{\dagger}\,e^{+i(\omega_k t - \mathbf{k}\cdot\mathbf{x})}]\,,$$

$$\omega_k \equiv \sqrt{k^2 + m^2}, \tag{3.63}$$

where the operator $\hat{b}_{\mathbf{k}}$ is unrelated to $\hat{a}_{\mathbf{k}}$, as a consequence of the fact that $\phi(x)$ is complex, $\phi(x) \neq \phi^{\dagger}(x)$.

By noting that the momentum conjugate to $\phi(x)$ is

$$\pi_{\phi}(x) = \frac{\partial\mathcal{L}}{\partial\dot{\phi}}(x) = \dot{\phi}^*(x) \tag{3.64}$$

and that $\pi_{\phi^*}(x) = \dot{\phi}(x)$, we then derive the following commutation relations for the $\hat{a}_{\mathbf{k}}^{(\dagger)}$ and $\hat{b}_{\mathbf{k}}^{(\dagger)}$ operators

$$[\hat{a}_{\mathbf{k}},\,\hat{a}_{\mathbf{k}'}^{\dagger}] = [\hat{b}_{\mathbf{k}},\,\hat{b}_{\mathbf{k}'}^{\dagger}] = (2\pi)^3\delta^{(3)}(\mathbf{k} - \mathbf{k}')\,, \tag{3.65}$$

with all other commutators vanishing. This implies that the theory has two kind of states: the "a-type" ones created by the $\hat{a}^{\dagger}(\mathbf{k})$ operator, and the "b-type" excitations created by $\hat{b}^{\dagger}(\mathbf{k})$. A (by now) straighforward calculation shows that both the a-type and the b-type excitations have energy given by the standard formula

$$H = \int \frac{d^3q}{(2\pi)^3}\,E_q[\hat{a}_{\mathbf{q}}^{\dagger}\,\hat{a}_{\mathbf{q}} + \hat{b}_{\mathbf{q}}^{\dagger}\,\hat{b}_{\mathbf{q}}] + \text{constant}\,. \tag{3.66}$$

What is the difference between a-type and the b-type excitations? To see this, let us discuss the symmetry of the Lagrangian under $\phi \mapsto e^{i\alpha}\,\phi$ (known as a "$U(1)$ symmetry" as the set of numbers of the form $e^{i\alpha}$, together with multiplication, forms the so-called $U(1)$ group). By taking $\alpha \ll 1$ we see that the invariance of the action under the $U(1)$ symmetry is equivalent to the symmetry in equation (3.48), where, if we denote $\phi_1 = \phi$ and $\phi_2 = \phi^*$, we have $F_{11} = i\phi$ and $F_{21} = -i\phi^*$. Then, the Noether theorem tells us that there will be one conserved current that, using equation (3.52), takes the form

$$j^{\mu} = i[(\partial^{\mu}\phi^*)\,\phi - \phi^*(\partial^{\mu}\phi)]\,, \tag{3.67}$$

so that the charge

$$Q \equiv \int d^3x \, j^0 \tag{3.68}$$

is constant.

By inserting the decomposition in equation (3.63) into equation (3.68) we obtain, after some algebra,

$$Q = \int \frac{d^3q}{(2\pi)^3} \, (\hat{a}_{\mathbf{q}}^\dagger \hat{a}_{\mathbf{q}} - \hat{b}_{\mathbf{q}}^\dagger \hat{b}_{\mathbf{q}} + (2\pi)^3 \, \delta_{\mathbf{q}}^{(3)}(0)) \, . \tag{3.69}$$

We thus discover that, while a-type and b-type particles both contribute positively to energy, they give contributions with opposite signs to Q. For this reason from now on we will refer, as is customary, to a-type excitations as "particles" and to b-type excitations as "antiparticles." The consequence is that the total number of particles minus antiparticles is a constant (as in the case of energy, the term proportional to $\delta_{\mathbf{q}}^{(3)}(0)$ in equation (3.69) is an irrelevant constant, and we will neglect it from now on).

To sum up, now we have two classes of one-excitation states: particles $|\varphi(\mathbf{p})\rangle \equiv \hat{a}_{\mathbf{p}}^\dagger |0\rangle$ and antiparticles $|\bar{\varphi}(\mathbf{p})\rangle \equiv \hat{b}_{\mathbf{p}}^\dagger |0\rangle$, which are eigenstates of H and Q with the following eigenvalues

$$H|\varphi(\mathbf{p})\rangle = E_p |\varphi(\mathbf{p})\rangle \, , \qquad H|\bar{\varphi}(\mathbf{p})\rangle = E_p |\bar{\varphi}(\mathbf{p})\rangle \, ,$$
$$Q|\varphi(\mathbf{p})\rangle = +1 \, |\varphi(\mathbf{p})\rangle \, , \qquad Q|\bar{\varphi}(\mathbf{p})\rangle = -1 \, |\bar{\varphi}(\mathbf{p})\rangle \, . \tag{3.70}$$

3.10 The nonrelativistic limit

The use of fields satisfying relativistic wave equations should not obscure the fact that these fields also describe nonrelativistic particles in the low energy limit. Indeed, it is often useful to take an explicit nonrelativistic limit in a field theory in situations where the antiparticle degrees of freedom are not particularly relevant. Let us see how to do this at the level of the field operator.

We can obtain the Schrödinger equation as the low energy limit of the Klein-Gordon equation by a field redefinition

$$\phi(t, x) = N e^{-imt} \Psi(t, x) \, , \tag{3.71}$$

where N is a normalization factor that we can choose later for convenience. The relativistic equation then turns into

$$\left(-im + \frac{\partial}{\partial t} \right)^2 \Psi - \nabla^2 \Psi + m^2 \Psi = 0 \tag{3.72}$$

or

$$-i\frac{\partial}{\partial t}\Psi - \frac{\nabla^2}{2m}\Psi + \frac{1}{2m}\left(\frac{\partial}{\partial t} \right)^2 \Psi = 0 \tag{3.73}$$

without any approximation. The nonrelativistic limit comes when the remaining time dependence in $\Psi(t)$ is small compared to the leading e^{-imt} factor so that the last term is small. Dropping it allows us to easily rewrite the equation as the free Schrödinger equation.

In the quantum field operator, we can make a similar transformation. Let us use the complex scalar field, recalling that it describes two degrees of freedom, the particle and its antiparticle. These can be separately treated in the field operator by defining

$$\phi(t, \mathbf{x}) = e^{-imt}\psi(t, \mathbf{x}) + \chi^\dagger(t, \mathbf{x}) \tag{3.74}$$

with

$$\psi(t, \mathbf{x}) = N \int \frac{d^3k}{(2\pi)^3\sqrt{2E_k}} e^{-i(E_k - m)t} e^{i\mathbf{k}\cdot\mathbf{x}} \hat{a}_\mathbf{k}. \tag{3.75}$$

Here χ is the antiparticle degree of freedom. If we consider external states that consist only of the particle states at low energy, then energy conservation tells us that the antiparticle cannot be directly produced and can exist only in intermediate states. In such a situation we can write the Lagrangian using only ψ, with interaction terms that can include the indirect effects of the antiparticle.

We should note that the choice of particle versus antiparticle is arbitrary. Is the electron the "particle" and the positron its "antiparticle," or should we use the positron as the "particle" and the electron as its "antiparticle"? The distinction does not matter physically. If we wanted to describe a collection of nonrelativistic positrons (and no electrons), we would label the positron as the particle. It would then require a lot of energy to excite an electron. In our world we have more nonrelativistic electrons, so we do the reverse. Mathematically, we would just replace ϕ by ϕ^* as the fundamental field. The Hermitian conjugation of the complex scalar field reverses the roles of the creation operators $\hat{a}_\mathbf{k}$ and $\hat{b}_\mathbf{k}$, which are the two degrees of freedom. But the Lagrangian involves only the combination $\phi^*\phi$, so either variable works equally well.

3.11 Photons

Special Relativity was first uncovered in classical electrodynamics, and the electromagnetic field has a beautiful relativistic description. We assume that you have seen this before, and we will quickly review it for notational purposes. The scalar potential $\varphi(x)$ and vector potential $\mathbf{A}(x)$ fit together into a four-vector $A^\mu = (\varphi, \mathbf{A})$. When packaged in a field strength tensor $F_{\mu\nu} = \partial_\mu A_\nu - \partial_\nu A_\mu$, Maxwell's equations $\partial^\mu F_{\mu\nu} = J_\nu$ emerge as the equations of motion for the Lagrangian

$$\mathcal{L} = -\frac{1}{4} F_{\mu\nu}F^{\mu\nu} - A_\mu J^\mu, \tag{3.76}$$

where $J_\mu(x)$ is the electromagnetic current density.

An important key to the construction of electromagnetism is gauge invariance.

The concept of gauge invariance sometimes comes across as obscure and somehow arbitrary. To demystify it a bit, consider, to start with, a theory with two scalar

From now on we will just show the contractions and the calculation behind it will be implied.

This construction allows us to construct all the states with all possible values of the energy. Because these are bosons, there can be more than one particle in a given energy state. For example, the normalized state

$$|n_1, 3\, n_2, n_3\rangle = \frac{1}{\sqrt{3!}} \hat{a}^\dagger_{n_3} \hat{a}^\dagger_{n_2} \hat{a}^\dagger_{n_2} \hat{a}^\dagger_{n_2} \hat{a}^\dagger_{n_1} |0\rangle \qquad (2.48)$$

has energy

$$H|n_1, 3\, n_2, n_3\rangle = (\hbar\,\omega_1 + 3\,\hbar\,\omega_2 + \hbar\,\omega_3)\,|n_1, 3\, n_2, n_3\rangle . \qquad (2.49)$$

We have finally arrived back to 1905 with quanta with the correct energy-frequency relation.

Our pathway to this point has been somewhat formal, in the sense that we have used the standard formalism for both classical mechanics and quantum mechanics. We just "turned the crank" and watched what emerged. The end product is quite intuitive. States are filled with quanta with the correct energy, and each quantum corresponds to a field that solves the wave equation. The operator character of fields, which seems nonintuitive at the start, is not so scary because it just turns into a number operator that counts the fields.[5] A different pedagogic approach would be to start with the existence of quanta and work from there to the idea of quantized fields. This is now easy to do—you are invited to read this chapter backward![6] Nevertheless, proceeding the way that we have done reinforces the point that the idea of "quanta of a field" is not a separate hypothesis from the basic postulates of quantum mechanics. It follows uniquely from the standard procedures of classical and quantum physics.

Let us recap what we have done here, because, perhaps without noticing, we have accomplished something extremely deep. We started from a discrete set of many particles and took the continuum limit. When quantizing the system in its continuum limit, we came up with a discrete set of states (such as the state $|n_1, 3\, n_2, n_3\rangle$ discussed in equation (2.48)). Each of these states is associated with what we call a set of "elementary particles"! Two main points here need to be stressed. The first point is that these "particles" have nothing to do with the original particles that make up the string. Actually, a lot of information has been lost by taking the continuum limit. (We started from a model with three parameters m, a, and k and ended up with a model with a single parameter v, so information has clearly been lost in the process.) The "elementary particles" found by quantization have thus very little to do with the "actual" particles that make up, at a microscopic level, our system. The second point is that Quantum Field Theory teaches us that we should not think as much in terms of individual particles as we should think in terms of excitations

---

[5] Later we will see that the operator also gives the correct counting factors for transitions due to interactions.

[6] Seriously, it is a good exercise to map out how you would explain to a novice the idea of a quantum field starting from the experimental evidence of quanta with $E = \hbar\,\omega$.

fields $\psi_1(x)$ and $\psi_2(x)$ whose Lagrangian, however, depends only on the combination $\psi_1(x) + \psi_2(x)$. This theory is invariant under the transformation $\psi_1(x) \to \psi_1(x) + \chi(x)$, $\psi_2(x) \to \psi_2(x) - \chi(x)$ for any arbitrary function $\chi(x)$, which is probably the simplest example of gauge invariance. Clearly, this is really the theory of *a single* scalar field $\psi(x) \equiv \psi_1(x) + \psi_2(x)$. So what we formulated as a theory of two scalars really features a single degree of freedom. A more rigorous way to see that this is the theory of a single degree of freedom is to notice that we can derive the Euler-Lagrange equations for any field by choosing small but otherwise arbitrary variations of the fields in the action. However, a variation given by a gauge transformation leads to a trivial Euler-Lagrange equation $0 = 0$ because by definition the variation of the action vanished under a gauge transformation. So gauge invariance is always telling us that we are formulating the theory in a number of degrees of freedom that is larger that its *actual* number of degrees of freedom.

In the case of electromagnetism, Maxwell's equations are invariant under the field change,

$$A_\mu(x) \to A'_\mu(x) = A_\mu(x) + \partial_\mu \chi(x) \tag{3.77}$$

for[1] *any* function $\chi(x)$, because the field strength tensor is unchanged: $F'_{\mu\nu} = F_{\mu\nu}$.

In the classical theory, gauge invariance is used to identify the propagating modes. The use of Lorentz gauge $\partial^\mu A_\mu = 0$ leads to the free field equation of motion

$$\Box A_\mu = 0. \tag{3.78}$$

When there is no source, $J^\mu = 0$, there is an additional gauge freedom to go to the *radiation gauge* with $A_0 = 0$, leaving two propagating transverse waves.

In the quantum theory, the equation of motion tells us that the photon is massless. In radiation gauge we can provide a representation for the photon field as

$$A_\mu(x) = \sum_{\lambda = \pm 1} \int \frac{d^3 p}{(2\pi)^3} \frac{1}{\sqrt{2\omega_p}} [\hat{a}_{\mathbf{p},\lambda}\, \epsilon_\mu(\hat{\mathbf{p}}, \lambda)\, e^{-ip\cdot x} + \hat{a}^\dagger_{\mathbf{p},\lambda}\, \epsilon^*_\mu(\hat{\mathbf{p}}, \lambda)\, e^{ip\cdot x}], \tag{3.79}$$

where $\hat{\mathbf{p}} = \mathbf{p}/p$. Masslessness is associated to the dispersion relation by $\omega_p = |\mathbf{p}|$, and radiation gauge tells us that the polarization vectors satisfy $p^\mu \epsilon_\mu(\hat{\mathbf{p}}, \lambda) = 0$, $\epsilon_0 = 0$. We normalize the polarization vectors to $\epsilon^*_\mu(\hat{\mathbf{p}}, \lambda)\epsilon^\mu(\hat{\mathbf{p}}, \lambda') = \delta_{\lambda,\lambda'}$. The choice of circular polarization with

$$\epsilon_\mu(p, \pm 1) = \frac{1}{\sqrt{2}}(0, 1, \pm i, 0) \tag{3.80}$$

for motion in the z direction turns out to correspond to photons with helicity ± 1.

---

[1]In the case of the two scalar fields $\psi_1(x)$ and $\psi_2(x)$ it is easy to "fix the gauge" and work just with $\psi(x) \equiv \psi_1(x) + \psi_2(x)$ and ignore the complications of a gauge theory. In the case of electromagnetism, however, it is impossible to fix the gauge while keeping Lorentz invariance. For this reason, people usually prefer to formulate the theory as a Lorentz-invariant one, even if it forces them to deal with the subtleties of gauge at various stages.

If we choose the commutation relations $[\hat{a}_{\mathbf{p},\lambda}, \hat{a}^{\dagger}_{\mathbf{p}',\lambda'}] = (2\pi)^3 \delta^{(3)}(\mathbf{p} - \mathbf{p}') \delta_{\lambda,\lambda'}$, we find the quanta of the electromagnetic field with $E = \hbar \omega_p$ from the Hamiltonian

$$H = \int d^3x \frac{1}{2} \left(\mathbf{E}^2 + \mathbf{B}^2 \right) = \sum_\lambda \int \frac{d^3p}{(2\pi)^3} \hbar \omega_p \, \hat{a}^{\dagger}_{\mathbf{p},\lambda} \hat{a}_{\mathbf{p},\lambda} , \qquad (3.81)$$

where we have dropped the zero-point energy. This procedure defines photons, which are quanta of the electromagnetic field.

3.12 Fermions—Preliminary

You can skip this section if you want.

At this point in the development of the subject we logically could present the formalism for Dirac fermions. However the technology for describing fermions is not needed to present the basic ideas of Quantum Field Theory, which is our mission in this book. Therefore our choice has been to defer the description of that technology until after we have developed the rest of the basics. So you can proceed without studying the details of relativistic fermions. If you prefer to encounter fermions in detail at this point of the development proceed to section 9.1.1, which provides an explanation of the subject.

We will continue the development of the subject using fields without spin. However, at a few points we will insert a comment or two about the results that occur with fermions. These are included for the benefit of those readers who have read section 9.1.1 first or who are returning to the book after completely reading it.

If you are continuing without a full study of Dirac fermions, here is a one-paragraph summary. The Dirac equation is a matrix equation. There are four solutions to the Dirac equation; those solutions describe a particle with spin up and spin down and an antiparticle with spin up and spin down. These solutions are described by four-component solutions related to the matrix structure of the Dirac equation: $u_\lambda(\mathbf{p})$ for the two particle solutions (λ here denotes the spin variable) and $v_\lambda(\mathbf{p})$ for the two antiparticle solutions. The complex conjugate of these solutions are written using the notation $\bar{u}_\lambda(\mathbf{p})$ and $\bar{v}_\lambda(\mathbf{p})$. These solutions are called Dirac spinors. States are described by using creation operators, which for fermions anticommute rather than commute. When taking matrix elements, the appropriate spinors will also appear, in a way that we will describe in section 9.1.1.

3.13 Why equal-time commutators?

The field theory commutation rules are taken at equal time. Why is this? At first it seems like a change from what we normally do in quantum mechanics when we use $[\hat{x}, \hat{p}] = i\hbar$. In fact it is fully consistent and the logic is the same, but the difference has to do with the difference between the Schrödinger picture and the Heisenberg picture.

Our quantum training tends to start using what is called the *Schrödinger picture*. Here the states have time dependence, $|\psi(t)\rangle$, and are evolved with the Hamiltonian

$$H|\psi_S(t)\rangle = i\hbar \frac{d}{dt}|\psi_S(t)\rangle, \tag{3.82}$$

which implies

$$|\psi_S(t)\rangle = e^{-\frac{iHt}{\hbar}}|\psi_S(0)\rangle. \tag{3.83}$$

The operators do not carry any time dependence. We can temporarily highlight this description by including the subscript S, such as

$$\hat{x}_S, \ \hat{p}_S, \ \hat{O}_S, \tag{3.84}$$

where $\hat{O}_S$ is an arbitrary operator in the theory. It is in this picture that we have time-independent commutators

$$[\hat{x}_S, \hat{p}_S] = i\hbar. \tag{3.85}$$

By contrast, in the *Heisenberg picture* the states do not have any time dependence, but the operators do. This can be accomplished by a unitary transformation

$$|\psi_H\rangle = e^{\frac{iHt}{\hbar}}|\psi_S(t)\rangle \quad \text{such that} \quad \frac{d}{dt}|\psi_H\rangle = 0. \tag{3.86}$$

For the operator matrix elements to remain the same, we must impose

$$\langle\psi_S(t)|\hat{O}_S|\psi'_S(t)\rangle = \langle\psi_S(0)|e^{\frac{iHt}{\hbar}}\hat{O}_S e^{-\frac{iHt}{\hbar}}|\psi'_S(0)\rangle = \langle\psi_H|\hat{O}_H(t)|\psi'_H\rangle, \tag{3.87}$$

where we have identified

$$\hat{O}_H(t) = e^{\frac{iHt}{\hbar}}\hat{O}_S e^{-\frac{iHt}{\hbar}}. \tag{3.88}$$

You might not have realized that in the Heisenberg picture one requires equal-time commutation rules. Indeed, at equal times we have

$$\begin{aligned}[\hat{x}_H(t), \hat{p}_H(t)] &= [e^{\frac{iHt}{\hbar}}\hat{x}_S e^{-\frac{iHt}{\hbar}}, e^{\frac{iHt}{\hbar}}\hat{p}_S e^{-\frac{iHt}{\hbar}}] \\ &= e^{\frac{iHt}{\hbar}}[\hat{x}_S, \hat{p}_S]e^{\frac{-iHt}{\hbar}} = i\hbar,\end{aligned} \tag{3.89}$$

and we can easily see that the non–equal-time commutator $[\hat{x}_H(t'), \hat{p}_H(t)]$ would *not* give the same result when $t' \neq t$. Note that operators that commute with the Hamiltonian, including the Hamiltonian itself, are also time independent in the Heisenberg picture.

We have implicitly used the Heisenberg picture in our development. The field operators $\phi(t, \mathbf{x})$ carry both position and time labels. This is how we think of classical fields, such as the electromagnetic fields, and it is natural to carry this over into the quantum theory. Moreover, this is required in a relativistic theory, in which Lorentz transformations involve both the position and time. We have also described our states by the quanta that they carry: $|\mathbf{p}\rangle = \hat{a}_\mathbf{p}^\dagger|0\rangle$. In the absence of interactions these states are time independent. So the Heisenberg picture is a natural one for quantum fields.

We will soon start to describe the interactions of fields. In this case, we invoke a third picture, the *interaction picture*. Here the full Hamiltonian is broken into two parts, the free field Hamiltonian and the interaction Hamiltonian:

$$H = H_0 + H_I. \tag{3.90}$$

The logic is that the free field Hamiltonian describes the free quanta of the theory. This is the part we focused on in chapter 2. If there are no interactions, these quanta determine the eigenstates of the theory, and these states do not change in time, just like in the Heisenberg picture. However, if there *are* interactions, the free quanta are not the eigenstates of the full theory, and they can then change in time. In an operator picture, this is accomplished by using H_0 first to define the states

$$|\psi_I(t)\rangle = e^{\frac{iH_0 t}{\hbar}} |\psi_S(t)\rangle, \tag{3.91}$$

with the corresponding operators

$$\hat{\mathcal{O}}_I(t) = e^{\frac{iH_0 t}{\hbar}} \hat{\mathcal{O}}_S e^{-\frac{iH_0 t}{\hbar}}. \tag{3.92}$$

The states will then evolve due to the presence of interactions

$$i\hbar\frac{d}{dt}|\psi_I(t)\rangle = e^{\frac{iH_0 t}{\hbar}} H_I e^{\frac{-iH_0 t}{\hbar}} |\psi_I(t)\rangle = H_I(t)|\psi_I(t)\rangle. \tag{3.93}$$

This equation will be solved by a unitary operator that evolves the states, so that

$$|\psi_I(t)\rangle = U_I(t, t_0) |\psi_I(t_0)\rangle, \tag{3.94}$$

where the *time-development operator* $U_I(t, t_0)$ will be the unit operator in the absence of H_I. We turn to the description of interactions in chapter 4.

Chapter summary: We accomplished a lot in this chapter. We have introduced a variety of fields, from nonrelativistic fields to photons, which we will continue to encounter throughout the book. There were important results, such as the general action principle, the rules for energy and momentum, and Noether's theorem. And there were some notational advances, including the continuum limit, the use of relativistic notation, and units that allow the key physics to be more visible. But still, we have only been treating noninteracting fields. It is time for a little action through a discussion of interactions to which we now turn.

CHAPTER 4

Interactions

So far, we have determined the exact eigenstates of simple Lagrangians. With some extra effort, field theorists can do the same for fermions or even gravitons. However, this is a boring world. These states never change in time. All that we see (literally) that is interesting comes from the interactions in our theories. It is the interactions that account for the complexity of matter and for changes and reactions. Theories with interactions rarely allow for exact solutions, and perturbation theory has been a remarkably effective tool to explore the nature of interactions.

In many ancient cultures there were wise men and women who carried special knowledge and who could read the future in special ways, for example by examining the sky or sheep entrails. In quantum physics, field theorists can simply look at a Lagrangian and tell us that the Higgs boson will decay to b quarks or whether the proton will be stable. To the uninitiated, this may look like an exotic skill, but it is straightforward once you know what to look for. This chapter and chapter 5 are devoted to the skill of reading Lagrangians and determining what happens. More prosaically, this amounts to determining the Feynman rules of a theory. The first step is the construction and analysis of interaction Lagrangians.

4.1 Example: Phonons again

Interactions are described by terms in the Lagrangian beyond those that determine the free field equations of motion. These terms carry powers of the fields beyond quadratic order. We can explicitly construct such a term using the one-dimensional phonon example that we used at the start of chapter 1. There are some useful lessons to be learned by doing this.

Let us assume that the initial potential between the particles is not purely harmonic but also has a quartic component

$$V(y_{j+1} - y_j) = \frac{k}{2} (\delta y_{j+1} - \delta y_j)^2 + \frac{\ell}{4} (\delta y_{j+1} - \delta y_j)^4, \tag{4.1}$$

where ℓ is some positive coefficient. If we follow the same path for defining the Lagrangian for the fields we will find a new term

$$\sum_j \frac{\ell}{4} (\delta y_{j+1} - \delta y_j)^4 \rightarrow \int \frac{dx}{a} \frac{\ell}{4} \left(\frac{a}{\sqrt{ak}} \frac{\partial \phi}{\partial x} \right)^4 \equiv \int dx \frac{\lambda}{4} \left(\frac{\partial \phi}{\partial x} \right)^4, \qquad (4.2)$$

where we have defined

$$\lambda = \frac{\ell\, a}{k^2}. \qquad (4.3)$$

This leads to the overall Lagrangian

$$\mathcal{L} = \frac{1}{2\,v^2} \left(\frac{\partial \phi}{\partial t} \right)^2 - \frac{1}{2} \left(\frac{\partial \phi}{\partial x} \right)^2 - \frac{\lambda}{4} \left(\frac{\partial \phi}{\partial x} \right)^4 \qquad (4.4)$$

for the field ϕ. The last term is the *interaction Lagrangian* for this system. The fact that this interaction contains four derivatives means that it becomes small at long wavelengths, so that the free field action is a good approximation at low energies.

There are a couple of insights about interaction Lagrangians that we can get from this example. One is related to the fact that the interaction in this specific example involves only derivatives of the field. This is related to an underlying symmetry of the system of springs. The symmetry is a *shift symmetry* in which the whole system is shifted uniformly

$$y_j(t) \rightarrow y_j(t) + b \qquad \forall j, \qquad (4.5)$$

where b is a constant that is independent of time and position. The absolute position of the system does not matter, only the relative displacement of its elements does. At the field level this is a constant shift in the field

$$\phi(t, x) \rightarrow \phi(t, x) + c. \qquad (4.6)$$

This symmetry allows interactions such as $(\partial_x \phi)^4$ but forbids others such as ϕ^4.

The other observation is that the interaction removes what was an apparent Lorentz-like symmetry of the free field theory. The free Lagrangian is invariant under a Lorentz-like transformation of the x and t coordinates. This is most easily seen using relativistic notation with

$$x^\mu = (v\,t,\, x) \quad \text{and} \quad \partial_\mu = \left(\frac{1}{v} \partial_t,\, \partial_x \right), \quad \text{with} \quad \mu = 0, 1, \qquad (4.7)$$

such that

$$S_{\text{free}} = \int dt\, dx \left[\frac{1}{2\,v^2} \left(\frac{\partial \phi}{\partial t} \right)^2 - \frac{1}{2} \left(\frac{\partial \phi}{\partial x} \right)^2 \right] = \int d^2x\, \frac{1}{2} \partial^\mu \phi\, \partial_\mu \phi, \qquad (4.8)$$

which is a $1 + 1$–dimensional version of the free massless Klein-Gordon Lagrangian of section 3.8. However, this apparent symmetry was *not* a symmetry of the underlying system. That the interaction term breaks the apparent symmetry is expected. This is an example of an *accidental symmetry* or *emergent symmetry*, that is, a symmetry that emerges only when considering terms below a certain order in the Lagrangian and which is not a symmetry of the more fundamental theory.

The lesson from this example is that the symmetries of a theory often determine the interactions. They tell us which types of interactions are allowed. In addition, we see that we can use the interaction terms to look for violations of the apparent symmetries of the low energy field theory.

4.2 Taking matrix elements

Theories where the Lagrangian contain terms with more than two powers of the fields lead to nonlinear equations of motion, which are generally impossible to solve. In chapter 5 we will present a set of rules that will allow us to compute observables for these theories in a perturbative expansion. As we will show there, a quantity that can give all the required information for a vast set of processes is the *matrix element*, which we have quickly seen in its quantum-mechanical incarnation in section 3.1. Simplifying things a bit for now, the matrix element $\mathcal{M}$ is related to the interaction Hamiltonian $\langle f|H_I|i\rangle$. To convince you to trust us here, we remind you that, in the interaction representation, H_I is the operator that evolves the states in time, so that $H_I|i\rangle$ must be related to the final state obtained in the presence of the interaction H_I if the initial state was $|i\rangle$, and $|\langle f|H_I|i\rangle|^2$ will be related to the probability that the state $|i\rangle$, evolved by the Hamiltonian H_I, turns into the final state $|f\rangle$.

When taking matrix elements involving states of quanta, a field operator either removes a particle or creates a particle. This action leads to the matrix elements for various transitions and eventually is encoded into the Feynman rules. Our goal in this section is to learn how to identify these matrix elements. This will eventually become automatic.

In section 2.4, we worked out what happens when an annihilation operator acts on a state with a single quantum in it. Using the contraction rule from that section, and working in the infinite-box limit of section 3.2, we readily find that

$$\langle 0|\phi(t, \mathbf{x})|\mathbf{p}\rangle = N_p \, e^{-i(\omega_p t - \mathbf{p}\cdot\mathbf{x})} \tag{4.9}$$

and

$$\langle \mathbf{p}|\phi(t, \mathbf{x})|0\rangle = N_p \, e^{i(\omega_p t - \mathbf{p}\cdot\mathbf{x})}, \tag{4.10}$$

where $N_p = 1/\sqrt{2\omega_p}$. If we wish to describe a transition from one state to another, we could use (for $\mathbf{p} \neq \mathbf{p}'$)

$$\langle \mathbf{p}'|\phi(t, \mathbf{x})\phi(t, \mathbf{x})|\mathbf{p}\rangle = \langle \mathbf{p}'|\phi(t, \mathbf{x})\phi(t, \mathbf{x})|\mathbf{p}\rangle + \langle \mathbf{p}'|\phi(t, \mathbf{x})\phi(t, \mathbf{x})|\mathbf{p}\rangle$$

$$= 2\, N_p \, N_{p'} \, e^{-i(\omega_p t - \mathbf{p}\cdot\mathbf{x})} e^{+i(\omega_{p'} t - \mathbf{p}'\cdot\mathbf{x})}. \tag{4.11}$$

Note the counting factor of 2. Similar factors will appear repeatedly. Our last example would be one with three quanta. You can work out the matrix element

$$\langle \mathbf{p}'', \mathbf{p}' | \phi^3(t, \mathbf{x}) | \mathbf{p} \rangle = 3! \, N_p \, N_{p'} \, N_{p''} \, e^{-i(\omega_p t - \mathbf{p} \cdot \mathbf{x})} e^{+i(\omega_{p'} t - \mathbf{p}' \cdot \mathbf{x})} e^{+i(\omega_{p''} t - \mathbf{p}'' \cdot \mathbf{x})}, \quad (4.12)$$

noting, again, the counting factor in front of the result. These matrix elements will end up describing the bookkeeping of transitions once we introduce interactions into the theory.

In these examples, all the creation operators in the fields were acting on the external states. We will also encounter situations where the operators in one field can act on the ones in another field rather than on the external state. For example, the simplest matrix element of this kind is

$$\langle 0 | \phi(t', \mathbf{x}') \, \phi(t, \mathbf{x}) | 0 \rangle. \quad (4.13)$$

Because $a_\mathbf{p} | 0 \rangle = \langle 0 | a_\mathbf{p}^\dagger = 0$, the only nonzero component of this matrix element is the one where the creation operator in $\phi(t, \mathbf{x})$ creates a particle and the annihilation operator in $\phi(t', \mathbf{x}')$ removes it, returning us to the vacuum state. The contraction here yields

$$\langle 0 | \underbrace{\phi(t', \mathbf{x}') \phi(t, \mathbf{x})}_{} | 0 \rangle = \int \frac{d^3 p}{(2\pi)^3} \frac{1}{2\omega_p} e^{+i(\omega_p t' - \mathbf{p} \cdot \mathbf{x}')} e^{-i(\omega_p t - \mathbf{p} \cdot \mathbf{x})}, \quad (4.14)$$

where we have included the explicit form of the normalization factor. There is a summation over the modes here because each mode can separately be created and annihilated.

4.3 Interactions of scalar fields

In favorable situations the form of the interactions is forced to be something specific by the symmetries of the theory or some other internal consistency condition. But we are not there yet. Instead we will just invent some interactions to use as examples until we get familiar with calculations in Quantum Field Theory. This is not as unreasonable as it sounds. In practice, when confronted with a novel situation that you want to describe by a field theory, you would first make up a simple interaction that mimics the expected physics. To do this you add some new term to the free field Lagrangian that will describe this physics.

For example, let us imagine that we want to describe the emission and absorption of a particle, as we know happens with the photon. At this stage, we do not yet want to bring in the complexities of fermions or of vector particles such as the photon, so we will do this using scalars. Consider the following interaction

$$\mathcal{L}_I = -g \, a(x) \, \phi^*(x) \, \phi(x). \quad (4.15)$$

Here $\phi(x) = \phi(t, \mathbf{x})$ is a complex scalar of the form introduced in section 3.9. The combination $\phi^* \phi$ has the ability (among other things) to remove a particle in an initial state and create one in the final state, in other words, to make a transition. As we discussed in section 3.7, the fact that this theory has a $U(1)$ symmetry implies that the number of ϕ particles minus that of ϕ antiparticles is conserved—that is, like electromagnetism, this theory conserves electric charge. The field $a(x) = a(t, \mathbf{x})$ is a real scalar. Because it appears singly it will either annihilate a particle from the initial state (absorption) or create one in the final state (emission). So this interaction can signal the reaction $\phi \to \phi + a$ and related reactions. The coupling constant g in front describes the strength of the interactions, much like the electromagnetic charge e.

Another important example provides a simple way to describe a scattering process. The simplest reaction scatters two particles in the initial state into two in the final state. This tells us that we need four fields to make this transition. A Lagrangian that does this using a real scalar field is

$$\mathcal{L}_I = -\frac{\lambda}{4!} \phi^4(x), \tag{4.16}$$

where λ is the coupling constant. The 4! in the denominator is just a convention, and different authors use different conventions. The minus sign in front of the interaction is not a convention once we set $\lambda > 0$. It is required to be there so that the Hamiltonian following from this Lagrangian, via $\mathcal{H} = \pi \dot{\phi} - \mathcal{L}$, does not have run-away negative energies for large ϕ. A related interaction can be introduced for the complex scalar field, using

$$\mathcal{L}_I = -\frac{\lambda}{4} [\phi^*(x)\phi(x)]^2. \tag{4.17}$$

Even though the same symbol λ is used here and in equation (4.16), these are describing different applications. The $\lambda\phi^4$ interaction is ubiquitous in physics as it can be used to model almost any short-range interaction.

Finally, let us include a three-particle coupling of a real scalar field, with a Lagrangian

$$\mathcal{L}_I = -\frac{g}{6} \phi^3(x). \tag{4.18}$$

For our purposes, this is included largely as a simpler version of the three-particle coupling in equation (4.15). However, similar triple couplings are present in real theories, such as Yang-Mills gauge theories (which describe the weak and strong interactions in the Standard Model), and in the field–theoretical description of the general theory of relativity, so we will consider it in what follows. For real scalar fields, this cubic interaction would generally be supplemented by the $\lambda\phi^4$ interaction so that the full Hamiltonian remains bounded from below.

4.4 Dimensional analysis with fields

After a while, it becomes a habit for physicists to perform dimensional analysis any time they look at a formula. This is a good practice also in Quantum Field Theory. There also turns out to be important and nontrivial physics hidden in the dimensions of various factors in a Lagrangian, such as the diagnosis for renormalizability, which we will discuss later. So let us describe how the pros do dimensional analysis on a Lagrangian.

The use of natural units makes this task much simpler. As described in section 3.3, in this system all dimensionful objects carry a power of a single unit, which we can describe as a mass unit M. Energies and momenta are of order M^1. This implies that derivatives with respect to position and time also carry the same dimension $\partial_\mu \sim M^1$. A power of x or t carries the dimension M^{-1}.

The action is dimensionless in $\hbar = 1$ units. It is given by $\int d^4x\, \mathcal{L}$, so that the Lagrangian density $\mathcal{L}$ must carry dimension M^4 (notice that this statement depends on the dimensionality of the space we are working in!). The Lagrangian for a scalar field starts off with $\partial_\mu \phi \partial^\mu \phi$, so that the field ϕ must carry mass dimension M^1. Such a mass dimension also is apparent from the canonical commutation rules. This is the key to dimensional analysis with Lagrangians.[1]

The dimensionality of coupling constants in the interaction Lagrangians can be determined by counting up the powers associated with the fields. For example, in the emission and absorption Lagrangian of equation (4.15), as well as the ϕ^3 interaction of equation (4.18), there are three scalar fields involved, adding up to mass dimension M^3. To make up the difference with the required scaling of the Lagrangian, one must endow the couplings g in both cases with mass dimension M^1. (In the real-world applications of Quantum Electrodynamics and Quantum Chromodynamics, for which these interactions are intended as proxies, there is also a derivative in the interactions, and the coupling is then dimensionless.) In contrast, a similar exercise with the ϕ^4 interaction of equation (4.16) reveals that the coupling λ is dimensionless.

There can also be interactions with inverse powers of a mass scale. For instance,

$$\mathcal{L} = \kappa\, h(x)\, \partial_\mu \phi^*(x) \partial^\mu \phi(x) \tag{4.19}$$

could be a scalar proxy for the gravitational coupling to matter. Here the scalar field $h(x)$ plays the role of the graviton, the field that propagates the gravitational interactions. The two derivatives mimic the actual form of the graviton coupling, which grows with the square of the energy. Roughly speaking, this is because Einstein taught us that gravity responds to energy, and the scalar field Hamiltonian involves two derivatives. In this case the coupling κ has an inverse mass dimension,

$$\kappa \sim \frac{1}{M_P}, \tag{4.20}$$

where for the actual gravitational coupling M_P is the Planck mass, $M_P = 1/\sqrt{8\pi G}$, where G is Newton's constant.

One can carry out a similar analysis in other settings. For example, in the one-dimensional phonon system described at the beginning of this chapter, the action is—as always—dimensionless and because $S = \int d^2x\, \mathcal{L}$, the Lagrangian carries

---

[1] For future reference, in four spacetime dimensions the photon field A_μ also carries dimension M^1 and fermions carry $M^{3/2}$.

mass dimension M^2. The field ϕ is then dimensionless. Dimensional analysis then shows that the coefficient λ in the interaction term of equation (4.4) must carry mass dimension M^{-2},

$$\lambda \equiv \frac{1}{M_*^2}, \qquad (4.21)$$

for some M_*. We noted that the interaction becomes small at low frequency/energy. Dimensional analysis answers the question of what "low" means here: this interaction is suppressed for energies much smaller than M_*. This scale is representative of the microscopic structure of the system.

4.5 Some transitions

What do the interactions do? Here is where we start reading the Lagrangians.

As a first example, consider scattering with the $\lambda \phi^4$ interaction of equation (4.16). The interaction has the ability to make a transition from particles in states with four-momenta p_1 and p_2 into states with four-momenta p_3 and p_4, i.e.,

$$\langle p_3, p_4 \,|i\mathcal{L}_I(x)|\, p_1, p_2\rangle = \left\langle p_3, p_4 \left| -i\frac{\lambda}{4!}\phi^4(x) \right| p_1, p_2 \right\rangle, \qquad (4.22)$$

where we have included a factor of i for later convenience.[2] When we use the field decomposition, we can see that to accomplish this transition, the annihilation operators in $\phi^4(x)$ must remove particles from the initial states, and the creation operators generate particles in the final state. There is a combinatoric factor of 4!, which can be seen by having four ways to chose an annihilation operator from $\phi^4(x)$ to act on p_1, three remaining ways to choose one to act on p_2, two remaining ways to create p_3, and one to create p_4. After the creation/annihilation operators have done their job, we are left with the momentum wavefunctions and normalization factors,

$$\langle p_3, p_4|i\mathcal{L}_I(x)|p_1, p_2\rangle = N_1\, N_2\, N_3\, N_4\, e^{-i(p_1+p_2-p_3-p_4)\cdot x} \times \left(-i\frac{4!}{4!}\lambda\right), \qquad (4.23)$$

with

$$N_i = \frac{1}{\sqrt{2E_i}}. \qquad (4.24)$$

At this stage, energy and momentum are not necessarily conserved. That will eventually come when we integrate over x.

The normalization factors and the exponential play a role in subsequent derivations, but are not what we are looking for here. To focus on the essential element,

---

[2] As we will see in chapter 5, the factor of i comes from the fact that we are considering a first-order expansion of the time-development operator in the interaction representation, $e^{-i\int H_I dt} \simeq 1 - i\int d^4x\, \mathcal{H}_I = 1 + i\int d^4x\, \mathcal{L}_I$.

let us define the *matrix element* $\mathcal{M}$ for general initial and final states via[3]

$$\langle f|i\mathcal{L}_I(x)|i\rangle = \left(\Pi_j N_j\right) \times e^{-i(\sum p_i - \sum p_f)\cdot x} \times \left(-i\mathcal{M}(p_i, p_f)\right), \tag{4.25}$$

where p_i and p_f denote the four-momenta of the particles in the initial and final states, respectively. The $-i$ in $-i\mathcal{M}$ is, again, conventional. With these definitions, we express equation (4.23) simply as

$$\mathcal{M} = \lambda. \tag{4.26}$$

If we were studying the theory of the complex scalar field, using the Lagrangian of equation (4.17), we would have a similar matrix element but with a different counting factor. Recall that ϕ annihilates a particle and creates an antiparticle, whereas ϕ^* annihilates an antiparticle and creates a particle. For the scattering of the particles, we would find

$$\langle \phi(p_3)\,\phi(p_4)|i\mathcal{L}_I(x)|\phi(p_1)\,\phi(p_2)\rangle = N_1\,N_2\,N_3\,N_4\,e^{-i(p_1+p_2-p_3-p_4)\cdot x} \times \left(-i\,4\frac{\lambda}{4}\right), \tag{4.27}$$

as there are only two ways to annihilate $\phi(p_1)$ and one way to remove $\phi(p_2)$ and a corresponding factor of 2 in the final state. The scattering of a particle and antiparticle is in this case also possible:

$$\langle \phi(p_3)\,\bar{\phi}(p_4)|i\mathcal{L}_I(x)|\phi(p_1)\,\bar{\phi}(p_2)\rangle = N_1\,N_2\,N_3\,N_4\,e^{-i(p_1+p_2-p_3-p_4)\cdot x} \times (-i\lambda). \tag{4.28}$$

Both of these processes thus yield the matrix element

$$\mathcal{M} = \lambda. \tag{4.29}$$

One can also consider the emission and absorption processes that are described by the interaction Lagrangian of equation (4.15). Here there are a variety of processes to be considered. For example, the emission of an a particle, $\phi \rightarrow \phi + a$, and the absorption process $\phi + a \rightarrow \phi$ are clearly related to each other. There are also the same processes for the antiparticles, $\bar{\phi} \rightarrow \bar{\phi} + a$ and $\bar{\phi} + a \rightarrow \bar{\phi}$. A bit more exotic are transitions involving a particle and antiparticle such as $a \rightarrow \phi + \bar{\phi}$ and $\phi + \bar{\phi} \rightarrow a$. In each case, there is only one way to create and annihilate the corresponding particle, and so there is no extra counting factor. All of these processes are described by the simple matrix element

$$\mathcal{M} = g. \tag{4.30}$$

These matrix elements are the building blocks of the Feynman diagrams that will be described in chapter 5. The operator character of the fields has been used to do the bookkeeping needed to produce the matrix element, but the transition elements

---

[3]This definition of the matrix element is valid only to leading order in perturbation theory, and we will show in chapter 5 that in general it will have a (much!) more complicated form, but this is good enough for the discussion in this chapter.

themselves are just numbers governed by the strength of the interaction in the form of an overall coupling.

Before closing this section, let us note that, while the computation of matrix elements gives a measure of the probability of the process to happen, this is not the end of the story. Even if a process has a large value of $|\mathcal{M}|^2$, it might be forbidden if it violates energy-momentum conservation. For instance, the process $\phi \to \phi + a$ we have just mentioned is forbidden by energy conservation: in the rest frame of the initial ϕ particle, the final state has an energy given by that of a plus the energy of the final ϕ, which is thus strictly larger than the mass of ϕ, that is, than the initial energy of the system.

4.6 The Feynman propagator

In the examples of section 4.5, the field operators all acted on the particles in the initial or final states. We will also encounter situations where the creation or annihilation operators act on similar operators in another field within the matrix element rather than in an external state. We will see that, in practice, these internal contractions come in a special form, known as the *Feynman propagator*, defined as[4]

$$iD_F(x, x') = \langle 0| T \left(\phi(x') \, \phi(x) \right) |0\rangle \quad , \tag{4.31}$$

where the *time-ordering* operation T defined in such way that the earlier time always appears on the right

$$T\left(\phi(x')\,\phi(x)\right) = \Theta(t'-t)\,\phi(x')\,\phi(x) + \Theta(t-t')\,\phi(x)\,\phi(x') \tag{4.32}$$

with $\Theta(t)$ being the Heaviside step function: $\Theta(t>0) = 1$, $\Theta(t<0) = 0$.

The Feynman propagator has a number of useful properties. Let us review them.

The Feynman propagator is a Green's function. Given any differential operator $\mathcal{O}_{x^\mu}$ acting on the coordinates x^μ, a Green's function $G(x, x')$ is defined as a solution[5] to the equation

$$\mathcal{O}_{x^\mu} G(x, x') = \delta^{(4)}(x^\mu - x'^\mu). \tag{4.33}$$

In particular, we are interested in the case in which $\mathcal{O}_{x^\mu}$ is the Klein-Gordon operator. Let us now show that, if the field $\phi(x)$ satisfies the Klein-Gordon equation

---

[4] In some Quantum Field Theory books this time-ordered product is defined as D_F rather than iD_F.

[5] Because $\mathcal{O}_{x^\mu}$ contains derivatives, it will have multiple Green's functions, corresponding to various boundary conditions.

$$\left(\frac{\partial^2}{\partial t^2} - \nabla^2 + m^2\right)\phi(x) = 0\,, \tag{4.34}$$

then the Feynman propagator satisfies

$$\left(\frac{\partial^2}{\partial t^2} - \nabla_{\mathbf{x}}^2 + m^2\right) D_F(\mathbf{x}, t; \mathbf{x}', t') = -\delta(t' - t)\,\delta^{(3)}(\mathbf{x}' - \mathbf{x})\,. \tag{4.35}$$

To do so, we compute, keeping in mind that $\dot{\Theta}(t) = \delta(t)$,

$$\frac{\partial}{\partial t} T(\phi(x')\,\phi(x)) = [-\delta(t' - t)\,\phi(t', \mathbf{x}')\,\phi(t, \mathbf{x}) + \delta(t - t')\,\phi(t, \mathbf{x})\,\phi(t', \mathbf{x}')]$$

$$+ [\Theta(t' - t)\,\phi(t', \mathbf{x}')\,\dot{\phi}(t, \mathbf{x}) + \Theta(t - t')\,\dot{\phi}(t, \mathbf{x})\,\phi(t', \mathbf{x}')]\,, \tag{4.36}$$

where the term in the first square brackets vanishes because, thanks to the Dirac delta functions, it is proportional to the equal-time commutator $[\phi(\mathbf{x}, t), \phi(\mathbf{x}', t)] = 0$. The second time derivative thus reads

$$\frac{\partial^2}{\partial t^2} T(\phi(x')\,\phi(x)) = [-\delta(t' - t)\,\phi(t', \mathbf{x}')\,\dot{\phi}(t, \mathbf{x}) + \delta(t - t')\,\dot{\phi}(t, \mathbf{x})\,\phi(t', \mathbf{x}')]$$

$$+ [\Theta(t' - t)\,\phi(t', \mathbf{x}')\,\ddot{\phi}(t, \mathbf{x}) + \Theta(t - t')\,\ddot{\phi}(t, \mathbf{x})\,\phi(t', \mathbf{x}')]\,. \tag{4.37}$$

Again, the term in the first square brackets can be simplified using the Dirac delta functions as well as the canonical quantization condition $[\phi(t, \mathbf{x}), \dot{\phi}(t, \mathbf{x}')] = i\delta^{(3)}(\mathbf{x}' - \mathbf{x})$. Moreover, the last line of equation (4.37), using the equation of motion, equation (4.34), turns out to be equivalent to $(\nabla_{\mathbf{x}}^2 - m^2)\,T\,(\phi(x')\,\phi(x))$.

Putting everything together, we thus obtain the identity

$$\frac{\partial^2}{\partial t^2} T(\phi(x')\phi(x)) = -i\delta(t' - t)\,\delta^{(3)}(\mathbf{x}' - \mathbf{x}) + (\nabla_{\mathbf{x}}^2 - m^2)T(\phi(x')\phi(x))\,, \tag{4.38}$$

from which equation (4.35) follows.

Expressions for $D_F(x, x')$. We can calculate $D_F(x, x')$ directly from its definition

$$iD_F(x, x') = \Theta(t' - t)\,\langle 0|\phi(x')\,\phi(x)|0\rangle + \Theta(t - t')\,\langle 0|\phi(x)\,\phi(x')|0\rangle\,, \tag{4.39}$$

using the decomposition of the field $\phi(x)$ into creation/annihilation operators that show that

$$\langle 0|\phi(x')\,\phi(x)|0\rangle = \int \frac{d^3p}{(2\pi)^3}\,\frac{e^{ip\cdot(x-x')}}{2\,\omega_p}\,, \tag{4.40}$$

so that

$$iD_F(x, x') = \int \frac{d^3p}{(2\pi)^3}\,\frac{1}{2\,\omega_p}\,(\Theta(t' - t)\,e^{+ip\cdot(x-x')} + \Theta(t - t')\,e^{-ip\cdot(x-x')})\,. \tag{4.41}$$

In this form, the energy is given by $p_0 = \omega_p = \sqrt{\mathbf{p}^2 + m^2}$, which is said to be "on-shell," satisfying $p_0^2 - \mathbf{p}^2 = m^2$.

Equation (4.41) shows explicitly that the Feynman propagator depends only on the difference $x'^\mu - x^\mu$, $D_F(x, x') = D_F(x - x')$. We can use this property as well as the fact that the propagator is a Green's function for the Klein-Gordon operator to find an expression of D_F that is much more useful than that in equation (4.41).

Equation (4.35), when written in terms of the Fourier transform

$$D_F(q) \equiv \int d^4x \, e^{iq \cdot (x-x')} D_F(x - x') \,, \tag{4.42}$$

takes the simple form $(-q^2 + m^2)iD_F(q) = -i$, so that we can write its solution as

$$D_F(x - x') = \int \frac{d^3q}{(2\pi)^3} \int \frac{dq_0}{2\pi} \frac{e^{-iq \cdot (x-x')}}{q^2 - m^2} \,. \tag{4.43}$$

In this form, the energy variable q_0 is not fixed to be on-shell, but instead it is an integration variable that runs over the whole real line, $-\infty < q_0 < +\infty$.

Now, however, we have a problem: the integrand in dq_0 in equation (4.43) has poles at the locations $q_0 = \pm\sqrt{\mathbf{q}^2 + m^2}$. How can we go through those singularities? We will not go *through* them, but we will rather go *around* them! This brings in the need to explore "$i\epsilon$ physics." Adding $i\epsilon$ to various formulas, with ϵ being a positive infinitesimal, would seem to be a somewhat arbitrary and even magical, not to say illegal, thing to do. In fact the locations of $i\epsilon$ in formulas is of major physical importance because it defines how one does contour integration, which in turn defines the physical properties of various functions. In this case, we will consider the integral in dq_0 in the complex-q_0 plane. This way we can perform small deformations of the integration path that allow the path to go around the poles. There are four different ways, however, to go around those two poles: for each pole we can choose the integration path to take a small positive or a small negative imaginary component. Deforming the integration path is equivalent to giving a small imaginary component to the poles. The Feynman propagator is obtained by considering poles at $q_0 = \pm\sqrt{\mathbf{q}^2 + m^2 - i\epsilon}$, with $\epsilon \to 0^+$. This is equivalent to considering

$$D_F(x - x') = \int \frac{d^3q}{(2\pi)^3} \int \frac{dq_0}{2\pi} \frac{e^{-iq_0(t-t')+i\mathbf{q}(\mathbf{x}-\mathbf{x}')}}{q^2 - m^2 + i\epsilon} \,. \tag{4.44}$$

These poles are pictured in figure 4.1

Let us now show that this expression of the Feynman propagator is equivalent to equation (4.41), which was found with a direct calculation.

To begin, consider the case where $t > t'$. Because the integrand is proportional to $e^{-iq_0(t-t')}$, we can close the integration path with a large half-circle on the $\text{Im}\{q_0\} < 0$ half plane. This circle includes the pole at $q_0 = +\sqrt{\mathbf{q}^2 + m^2 - i\epsilon}$, and we can use

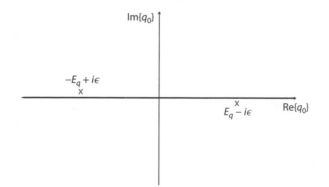

Figure 4.1. The poles, marked with "×" in the Feynman propagator in the complex-q_0 plane, including the factors of $i\epsilon$. Here $E_q = +\sqrt{\mathbf{q}^2 + m^2}$.

the residue theorem to evaluate the integral to

$$D_F(x - x')\Big|_{t>t'} = -i \int \frac{d^3q}{(2\pi)^3} \frac{e^{-i\sqrt{\mathbf{q}^2+m^2}(t-t')+i\mathbf{q}\cdot(\mathbf{x}-\mathbf{x}')}}{2\sqrt{\mathbf{q}^2+m^2}}. \tag{4.45}$$

An analogous calculation for the case $t < t'$, in which we must close the integration contour on the upper half of the complex-q_0 plane, leads, after changing sign to integration variable $\mathbf{q}$, to

$$D_F(x - x')\Big|_{t'>t} = -i \int \frac{d^3q}{(2\pi)^3} \frac{e^{-i\sqrt{\mathbf{q}^2+m^2}(t'-t)+i\mathbf{q}\cdot(\mathbf{x}'-\mathbf{x})}}{2\sqrt{\mathbf{q}^2+m^2}}. \tag{4.46}$$

The equivalence of equations (4.45) and (4.46) to equation (4.41) shows, finally, that equation (4.44) indeed provides the correct expression of the Feynman propagator.

Before concluding this section, let us note that a derivation analogous to that of equations (4.45) and (4.46) shows that

$$iD_{\text{ret}}(x - x') = \int \frac{d^3q}{(2\pi)^3} \int \frac{dq_0}{2\pi} \frac{e^{-iq_0(t-t')+i\mathbf{q}\cdot(\mathbf{x}-\mathbf{x}')}}{(q_0 + i\epsilon)^2 - \mathbf{q}^2 - m^2} \tag{4.47}$$

provides the *retarded* (also known as *causal*) propagator for the Klein-Gordon operator, which vanishes identically as long as $t' < t$. Because the propagators $D(x - x')$ are Green's functions, then any expression of the form

$$\varphi(x) = -\int d^4x' \, D(x - x') \, J(x') \tag{4.48}$$

is a solution of the Klein-Gordon equation with a source J, that is

$$\left(\frac{\partial^2}{\partial t^2} - \nabla^2 + m^2\right)\varphi(x) = J(x), \tag{4.49}$$

as it is easy to see by acting with the Klein-Gordon operator on both sides of equation (4.48) and using equation (4.35). In particular, this implies that if the source $J(x)$ vanishes before some given time t_0, $J(t < t_0, \mathbf{x}) = 0$, then the solution $-\int d^4x' \, D_{\text{ret}}(x - x') J(x')$ is also vanishing for $t < t_0$. On the other hand, $-\int d^4x' \, D_F(x - x') J(x')$ will not vanish for $t < t_0$ even if the source does vanish at early times. It is thus often said that while the retarded propagator propagates solutions to the future, the Feynman propagator propagates positive frequency modes (those proportional to $e^{-i\omega t}$) to the future and negative frequency modes ($\propto e^{i\omega t}$) to the past. This will make more sense when we see the use of the Feynman propagator in perturbation theory.

Chapter summary: We have seen how quantum fields can be used to characterize transitions between quantized states when there are several fields interacting in the Lagrangian. This is at the heart of Quantum Field Theory. Particles can be created and absorbed. Particles can change their identities. The matrix elements for such transitions are determined by the creation and annihilation operators contained in the quantum field operator, up to an overall coupling constant that gives the strength of the interaction. We have also had our first introduction to the Feynman propagator. This is a special matrix element that describes the creation and subsequent annihilation of a particle that does not exist in the initial or final state, but that does appear as an intermediate state. The matrix elements that we have learned to calculate here will soon be put together to make physical reactions.

CHAPTER 5

Feynman rules

There is a widespread folk tale about "stone soup." In the tale, a soup is started using only a stone and water. The cook describes the wonderful soup that will result once some garnishes are added (some carrots perhaps and potatoes and a bit of ham, etc.). The neighbors all contribute some of the extras. In the end, the stone comes out and is discarded and everyone enjoys the stone soup. This is what we are about to do. We will start off with the time-development operator, add some interactions and propagators, and create the field-theoretical rules. However, the time-development operator quickly gets discarded and most field theorists never mention it, simply using the rules that emerged.[1]

5.1 The time-development operator

Recall that we are living in the interaction picture (see section 3.13). The Hamiltonian is divided into a free field portion H_0 and an interaction Hamiltonian H_I. States do not change with time unless H_I is nonzero, and the basis states we will use are the free particle states $|\mathbf{p}\rangle$. States change in time due to the interaction picture *time-development operator* $U_I(t, t_0)$:

$$|\psi(t)\rangle = U_I(t, t_0) |\psi(t_0)\rangle . \tag{5.1}$$

Let us briefly review the construction of the time-development operator, which is treated the same way here as in quantum mechanics texts. The states satisfy the differential equation

$$i\frac{d}{dt}|\psi(t)\rangle = H_I(t) |\psi(t)\rangle , \tag{5.2}$$

which we can turn into an integral equation

$$|\psi(t)\rangle = |\psi(t_0)\rangle - i \int_{t_0}^{t} dt'\, H_I(t') |\psi(t')\rangle . \tag{5.3}$$

---

[1] In chapter 8 we will develop a soup of the same flavor using path integrals.

Now, we start by assuming that the effect of H_I is "small." (As often happens in physics, we assume at the start that something is small, and only at the end of the process will we find out for what values of the parameters such approximations are justified and what "small" really means.) In this case, we can solve equation (5.3) iteratively

$$|\psi(t)\rangle = |\psi(t_0)\rangle - i \int_{t_0}^{t} dt'\, H_I(t')\, |\psi(t_0)\rangle + (-i)^2 \int_{t_0}^{t} dt'\, H_I(t') \int_{t_0}^{t'} dt''\, H_I(t'')\, |\psi(t_0)\rangle$$

$$+ (-i)^3 \int_{t_0}^{t} dt'\, H_I(t') \int_{t_0}^{t'} dt''\, H_I(t'') \int_{t_0}^{t''} dt'''\, H_I(t''')|\psi(t_0)\rangle + \dots \ .$$

$$(5.4)$$

Note that the interactions here occur with the earliest interaction to the right ($t > t' > t'' > t''' > \dots$).

We can use the operation of *time ordering*, which we have already seen in section 4.6, to write

$$\int_{t_0}^{t} dt'\, H_I(t') \int_{t_0}^{t'} dt''\, H_I(t'') = \frac{1}{2} \int_{t_0}^{t} dt' \int_{t_0}^{t} dt''\, T\left(H_I(t')\, H_I(t'')\right),$$

$$\int_{t_0}^{t} dt'\, H_I(t') \int_{t_0}^{t'} dt''\, H_I(t'') \int_{t_0}^{t''} dt'''\, H_I(t''')$$

$$= \frac{1}{3!} \int_{t_0}^{t} dt' \int_{t_0}^{t} dt'' \int_{t_0}^{t} dt'''\, T\left(H_I(t')\, H_I(t'')\, H_I(t''')\right),$$

$$(5.5)$$

and so on, as it is easy to verify using the definition of the T-product, equation (4.32).

The resulting series can thus be expressed via the compact notation

$$\boxed{U_I(t_f, t_i) = T \exp\left[-i \int_{t_i}^{t_f} dt \int d^3x\, \mathcal{H}_I(t)\right]},$$

$$(5.6)$$

where the exponential is defined by its Taylor expansion, and the overall time-ordering operation T applies to each term in the series. The time integral in every integration runs the full range from t_i to t_f.

When we apply this time-development operator to our basis state, we imagine starting at very early times $t \to -\infty$ with some initial state, and we will denote it by $|i\rangle$. This state then gets evolved to a late time $t \to +\infty$, so that at late times it is given by $U_I(\infty, -\infty)|i\rangle$. The question we ask is: what is the amplitude for the system to end up into a different state, which we will call $|f\rangle$? This amplitude is denoted by

$$S_{fi} = \langle f | U_I(\infty, -\infty) | i \rangle \quad , \tag{5.7}$$

where S_{fi} is called the *S-matrix*, a (generally infinite-dimensional) matrix whose elements are the transition amplitudes from initial to final states. For a field theory, integrating over space and time implies

$$S = T \exp\left[-i \int d^4x \, \mathcal{H}_I \right] = T \exp\left[i \int d^4x \, \mathcal{L}_I \right]. \tag{5.8}$$

We further define the *transition matrix T* via

$$S = 1 - iT. \tag{5.9}$$

The unit operator does not generate any transitions, so all the interesting physics is in the T matrix.

As a first example let us consider the scattering of two quanta of ϕ field using the $\lambda \phi^4$ Lagrangian. Expanding to second order in λ, we obtain

$$S_{fi} = \left\langle p_4, p_3 \left| \left[1 - i \int d^4x \frac{\lambda}{4!} \phi^4(x) \right. \right. \right.$$
$$\left. \left. \left. - \frac{1}{2} \int d^4x \, d^4y \left(\frac{\lambda}{4!} \right)^2 T \left(\phi^4(x) \, \phi^4(y) \right) + O(\lambda^3) \right] \right| p_1, p_2 \right\rangle. \tag{5.10}$$

The term linear in λ does not need any time ordering because all the fields are evaluated at the same time, but this is not the case for the term proportional to λ^2. The linear term has just the right number of operators to make the transition— four creation/annihilation operators in $\phi^4(x)$ that we can use to cancel the two creation operators coming from the initial state and the two annihilation operators coming from the final state. We have calculated the matrix element of this term in section 4.5. The quadratic term has too many field operators for us at the moment, so we will return to it in section 5.4.

Using the matrix element calculated in section 4.5, we readily find, to leading order in λ,

$$-i\,T_{fi} = \int d^4x \frac{e^{-i(p_1+p_2-p_3-p_4)\cdot x}}{\sqrt{2E_1\,2E_2\,2E_3\,2E_4}}(-i\lambda) = (2\pi)^4 \frac{\delta^{(4)}(p_1+p_2-p_3-p_4)}{\sqrt{2E_1\,2E_2\,2E_3\,2E_4}}(-i\mathcal{M}),$$
$$\tag{5.11}$$

with $\mathcal{M} = \lambda + \mathcal{O}(\lambda^2)$. This structure is typical of the answers that we will get. There is always an overall energy-momentum–conserving delta function, coming from the exponentials that appear in the matrix elements. There are the ubiquitous normalization factors. And then there is a matrix element that contains the essential information of the theory, which in this case is the strength of scattering in the parameter λ. After a few more examples, we will get rid of the common factors, saving them for later, and just quote the matrix element.

5.2 Tree diagrams

The calculation of the transition matrix in equation (5.11) was quite straightforward. Let us now try something a bit more complicated. To do so, let us consider a generalization of the first example considered in section 4.3, a simplified model of electromagnetic interactions that does not deal with the complications of spin.

Our Lagrangian will contain two complex scalars ϕ_e and ϕ_p with masses m_e and m_p that will play the role of electron and proton, respectively, and a real scalar a that plays the role of the photon. The interaction Lagrangian will contain the photon-electron and the photon-proton interaction terms,

$$\mathcal{L}_I = -g_e \, a \, \phi_e^* \, \phi_e - g_p \, a \, \phi_p^* \, \phi_p \,. \tag{5.12}$$

Let us consider, in this simplified model of electromagnetism, the analog of the Rutherford scattering $ep \to ep$, where the incoming ϕ_e and ϕ_p have four-momenta p_1 and p_2, and the outgoing ϕ_e and ϕ_p have four-momenta p_3 and p_4, respectively. Because there are no quanta of the field a in the initial nor the final state, while $\mathcal{L}_I$ contains a to the first power, the first-order contribution to the S-matrix vanishes. We thus have to go to the second order. At that order, $\mathcal{L}_I^2$ contains $(\phi_e^* \, \phi_e)^2$, $(\phi_p^* \, \phi_p)^2$, and $(\phi_e^* \, \phi_e)(\phi_p^* \, \phi_p)$ terms. Because we are interested in processes in which both the e and the p fields participate to the interactions, we keep only the latter terms. To sum up, we get

$$-i\, T_{fi} = -\left\langle p_4, p_3 \,\middle|\, \int d^4x \, d^4y \, g_e \, g_p \, T\left(a(x)\phi_e^*(x)\phi_e(x) \, a(y)\phi_p^*(y)\phi_p(y)\right) \middle|\, p_1, p_2 \right\rangle, \tag{5.13}$$

where the factor of $1/2$ from the expansion of the time-development operator has been canceled by the factor of 2 coming from the double product.

Because ϕ_e and ϕ_e^* are computed at the same time, we can pull them out of the T-product, and the same holds for ϕ_p and ϕ_p^*. Remembering that $|p_1\rangle$ and $|p_3\rangle$ are states that contain quanta of ϕ_e, and that $|p_2\rangle$ and $|p_4\rangle$ contain quanta of ϕ_p, we can thus simplify the expression of T_{fi} to

$$T_{fi} = -i g_e \, g_p \int d^4x \, d^4y \langle p_3|\phi_e^*(x)\,\phi_e(x)|p_1\rangle \, \langle p_4|\phi_p^*(y)\,\phi_p(y)|p_2\rangle \, \langle 0|\, T\left(a(x)\,a(y)\right)|0\rangle, \tag{5.14}$$

where we recognize, in the last term, the Feynman propagator for the a-field. While we will provide a detailed derivation of Feynman diagrams in section 5.6, it is worth showing here the Feynman diagram for equation (5.13). We do this in figure 5.1.

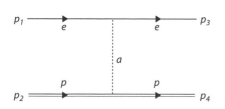

Figure 5.1. The Feynman diagram associated with the matrix element in equation (5.13). The external lines labeled $p_1, \ldots, p_4$ denote the initial and final ϕ_e (single solid line) and ϕ_p (double solid line) particles; the vertices correspond to the $a\phi_e^*\phi_e$ and $a\phi_p^*\phi_p$ interaction terms in the Lagrangian; and the dashed line corresponds to the a propagator.

The topology of the diagram indicates the ingredients we used to compute the matrix element for this process at this order in perturbation theory: one vertex takes care of eliminating one ϕ_e from the initial state and one from the final state; the other vertex does the same with the initial and final ϕ_p; and the propagator accounts for the pair of a field operators generated by those vertices. In section 5.6 we will see how this diagram can also be used to derive the mathematical expression of the matrix element $\mathcal{M}$.

Using the formulas found in chapter 4, we thus obtain

$$T_{fi} = -ig_e\,g_p \int d^4x\,d^4y \frac{e^{-i(p_1-p_3)\cdot x}}{\sqrt{2E_1\,2E_3}} \times \frac{e^{-i(p_2-p_4)\cdot y}}{\sqrt{2E_2\,2E_4}} \times i \int \frac{d^4q}{(2\pi)^4} \frac{e^{-iq\cdot(x-y)}}{q^2+i\epsilon}$$

$$= (2\pi)^4 \frac{\delta^{(4)}(p_1+p_2-p_3-p_4)}{\sqrt{2E_1\,2E_2\,2E_3\,2E_4}} \frac{g_e\,g_p}{(p_1-p_3)^2+i\epsilon}, \tag{5.15}$$

giving finally the simple result

$$\mathcal{M} = \frac{g_e\,g_p}{(p_1-p_3)^2+i\epsilon}. \tag{5.16}$$

While the "electrons" and the "protons" in this example are physical, and have momenta that satisfy the so-called *on-shell* condition $p^2 = m^2$, the "photon" a, even if it has a central role in our calculation, does not appear as a physical particle. We can tell this by observing that its four-momentum q^μ satisfies $q^2 = (p_1-p_3)^2 = 2m_e^2 - 2\sqrt{m_e^2+\mathbf{p}_1^2}\sqrt{m_e^2+\mathbf{p}_3^2} + 2\,\mathbf{p}_1\cdot\mathbf{p}_3$, which is different from the value ($q^2=0$) it would have if the *particle* were a physical state: this particle is *off-shell*. What is referred to as the "shell" (or, more precisely, the "mass shell") is the hyperbolic surface $p_0^2 = m^2 + \mathbf{p}^2$ in four-momentum space.

Before concluding this section, we note that the calculations we performed here are simple enough that we never needed to deal with any integral that was not trivially computed using a Dirac delta function or its Fourier transform. These calculations are said to be "at tree level," and the reason for this denomination will become clear in section 5.6, where we will discuss Feynman diagrams. More involved calculations can contain nontrivial integrals—these are said to be "at loop level"—and we will see one of them in detail in section 5.4.

5.3 Wick's theorem

Let us be a little more explicit about a result that we used in calculations in this chapter. When we had several fields within a time-ordered product, some of the fields were taken to act on the external states. The leftover ones were contained within the time-ordered product and therefore turned into propagators. We can use Wick's theorem to make this procedure more systematic.

To start we define a *normal-ordered product*: within the product of fields, all annihilation operators sit to the right of creation operators. This means that the annihilation operators must act only on external states and cannot act on any of the operators that are in the product of fields. The normal-ordered product is denoted by double dots on each side, for example, $:\phi \ldots \phi:$. As an example, here is the explicit form of the normal-ordered product of two fields:

$$
:\phi(x)\phi(y): = : \int \frac{d^3k}{(2\pi)^3} \frac{1}{\sqrt{2\omega_k}} \left[\hat{a}_{\mathbf{k}} e^{-ik\cdot x} + \hat{a}_{\mathbf{k}}^\dagger e^{ik\cdot x} \right]
$$

$$
\times \int \frac{d^3q}{(2\pi)^3} \frac{1}{\sqrt{2\omega_q}} \left[\hat{a}_{\mathbf{q}} e^{-iq\cdot y} + \hat{a}_{\mathbf{q}}^\dagger e^{+iq\cdot y} \right]:
$$

$$
= \int \frac{d^3k}{(2\pi)^3 \sqrt{2\omega_k}} \int \frac{d^3q}{(2\pi)^3 \sqrt{2\omega_q}}
$$

$$
\times \left(\hat{a}_{\mathbf{k}} \hat{a}_{\mathbf{q}} e^{-ik\cdot x - iq\cdot y} + \hat{a}_{\mathbf{q}}^\dagger \hat{a}_{\mathbf{k}} e^{-ik\cdot x + iq\cdot y} + \hat{a}_{\mathbf{k}}^\dagger \hat{a}_{\mathbf{q}} e^{ik\cdot x - iq\cdot y} + \hat{a}_{\mathbf{k}}^\dagger \hat{a}_{\mathbf{k}}^\dagger e^{ik\cdot x + iq\cdot y} \right). \quad (5.17)
$$

Of the four terms in the last line above, only the second one has been modified so that the annihilation operator is on the right of the creation operator.

Before going ahead with the formulation of Wick's theorem, let us note that normal ordering is precisely the operation that we perform on the Hamiltonian of the free field when we drop the zero-point energy:

$$
H = \int \frac{d^3p}{(2\pi)^3} \frac{\omega_p}{2} \left[\hat{a}_{\mathbf{p}}^\dagger \hat{a}_{\mathbf{p}} + \hat{a}_{\mathbf{p}} \hat{a}_{\mathbf{p}}^\dagger \right] \; \Rightarrow \; :\hat{H}: = \int \frac{d^3p}{(2\pi)^3} \omega_p \hat{a}_{\mathbf{p}}^\dagger \hat{a}_{\mathbf{p}}. \quad (5.18)
$$

The main property of the normal-ordered quantities is that their expectation value on the vacuum does vanish:

$$
\langle 0| :\phi \ldots \phi: |0\rangle = 0. \quad (5.19)
$$

Now, Wick's theorem relates the product of strings of field operators to the normal-ordered expression of the same string and combinations of Feynman propagators built with pairs of fields appearing in the string. The content is intuitive. Creation and annihilation operators either act on external states or on other fields in the time-ordered product. Those acting on other operators within the time-ordered product form propagators. Those acting on external states are the same as if they were in a normal-ordered product. One must consider all permutations of such possibilities.

The theorem can be proved formally by induction, but instead of giving a cumbersome general formula, let us write explicit examples. For a string of two scalar fields, Wick's theorem states that

$$T\left(\phi(x)\,\phi(y)\right) = :\phi(x)\,\phi(y): + \langle 0|T\left(\phi(x)\,\phi(y)\right)|0\rangle, \qquad (5.20)$$

for three fields,

$$T\left(\phi(x)\,\phi(y)\,\phi(z)\right) = :\phi(x)\,\phi(y)\,\phi(z): + :\phi(x): \langle 0|T\left(\phi(y)\,\phi(z)\right)|0\rangle$$

$$+ :\phi(y): \langle 0|T\left(\phi(x)\,\phi(z)\right)|0\rangle + :\phi(z): \langle 0|T\left(\phi(x)\,\phi(y)\right)|0\rangle,$$

$$(5.21)$$

and, for four fields,

$$T\left(\phi(x_1)\,\phi(x_2)\,\phi(x_3)\,\phi(x_4)\right) = :\phi(x_1)\,\phi(x_2)\,\phi(x_3)\,\phi(x_4):$$

$$+ \sum_{\text{perms}} :\phi(x_i)\,\phi(x_j): \langle 0|T\left(\phi(x_k)\,\phi(x_l)\right)|0\rangle$$

$$+ \sum_{\text{perms}} \langle 0|T\left(\phi(x_i)\,\phi(x_j)\right)|0\rangle\,\langle 0|T\left(\phi(x_k)\,\phi(x_l)\right)|0\rangle,$$

$$(5.22)$$

where "perms" denotes all the permutations of the set $\{1, 2, 3, 4\}$.

Because the expectation value of a normal product on the vacuum vanishes, Wick's theorem essentially states that all the operators that cannot be used to act on an external particle state must be paired to form Feynman propagators.

Let us now see how this works in a more complicated example, and let us also use this example to introduce Feynman diagrams.

5.4 Loops

Let us return to the calculation of the scattering amplitude in the $\lambda\phi^4$ theory. In section 4.5 we calculated the effect of the first-order term, in which the four fields were exactly the correct number necessary to remove two particles from the initial state and create two in the final state. If we look at the second-order term

$$-iT_{fi}^{(2)} = -\frac{1}{2}\left(\frac{\lambda}{4!}\right)^2 \int d^4x\,d^4y\,\langle p_3,\,p_4|\,T\left(\phi^4(x)\,\phi^4(y)\right)|p_1,\,p_2\rangle, \qquad (5.23)$$

we see that we have eight fields, which is more than we need to get rid of the creation/annihilation operators coming from the final/initial states. Wick's theorem tells us that the four fields that will not act on the external states will have to be contracted into propagators.

There are many possibilities for the action of the fields. To keep all the possibilities under control, let us start drawing the following diagram:

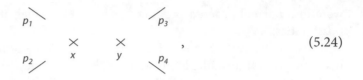

$$(5.24)$$

where each external *leg*, denoted by $p_1, \ldots, p_4$, corresponds to one creation/annihilation operator $\hat{a}^\dagger_{\mathbf{p}_1}, \ldots, \hat{a}_{\mathbf{p}_4}$, and the *vertices* labeled x and y have four legs each, corresponding to the fact that the $\phi(x)^4$ contain products of four creation/annihilation operators that either can be used to remove the external $\hat{a}^\dagger_{\mathbf{p}_1}, \ldots, \hat{a}_{\mathbf{p}_4}$ operators, or can be paired to other creation/annihilation operators present in $\phi(x)^4$ or in $\phi(y)^4$ to form Feynman propagators.

In this way, we can get several *Feynman diagrams* with different topologies. For each of them, we have to be careful to compute multiplicities. Let us see the different diagrams we can get.

To begin, we can connect two initial legs p_1 and p_2 to the x vertex, the two final legs to the y vertex, and then the free legs of the two vertices to each other:

$$(5.25)$$

There are two equivalent ways of labeling the vertices $(x \leftrightarrow y)$; four ways of connecting the p_1 leg to its vertex; and once one of the legs of that vertex is taken by p_1, three ways of connecting p_2 to the same vertex. Analogously, there are 4×3 ways of connecting the legs p_3 and p_4 to the other vertex. The operator $\hat{a}_\mathbf{q}$ used to eliminate $\hat{a}^\dagger_{\mathbf{p}_1}$ comes with a factor $e^{-iq \cdot x}/\sqrt{2\omega_q}$, where $q = p_1$ once $\hat{a}_\mathbf{q}$ acts on $\hat{a}^\dagger_{\mathbf{p}_1}$. Likewise, by acting on the leg p_2 we get a term $e^{-ip_2 \cdot x}/\sqrt{2\omega_{p_2}}$, whereas the final legs give terms $e^{ip_3 \cdot y}/\sqrt{2\omega_{p_3}}$ and $e^{ip_4 \cdot y}/\sqrt{2\omega_{p_4}}$.

Now that we have used all the field operators to eliminate the external legs, we are left with $\langle 0|T\left(\phi^2(x)\,\phi^2(y)\right)|0\rangle$ that can give $\langle 0|T\left(\phi(x)\,\phi(y)\right)|0\rangle\,\langle 0|T\left(\phi(x)\,\phi(y)\right)|0\rangle$ in two equivalent ways.

To sum up, the diagram with the topology in equation (5.25) evaluates to

$$T_{fi}^{(2)}[\text{Diagram (5.25)}] = -\frac{1}{2}\left(\frac{\lambda}{4!}\right)^2 \times 2 \times (4 \times 3)^2 \times 2$$

$$\times \int d^4x\, d^4y\, \frac{e^{-i(p_1+p_2)\cdot x}}{\sqrt{2\omega_{p_1}}\,\sqrt{2\omega_{p_2}}}\, \frac{e^{i(p_3+p_4)\cdot y}}{\sqrt{2\omega_{p_3}}\,\sqrt{2\omega_{p_4}}}$$

$$\times \int \frac{d^4q}{(2\pi)^4}\frac{e^{iq\cdot(x-y)}}{q^2-m^2+i\epsilon} \int \frac{d^4q'}{(2\pi)^4}\frac{e^{iq'\cdot(x-y)}}{q'^2-m^2+i\epsilon}$$

$$= -\frac{1}{2}\lambda^2\,(2\pi)^4\frac{\delta^{(4)}(p_1+p_2-p_3-p_4)}{\sqrt{2\omega_{p_1}\,2\omega_{p_2}\,2\omega_{p_3}\,2\omega_{p_4}}}\,I(p_1+p_2). \quad (5.26)$$

where we have defined

$$I(p) \equiv \int \frac{d^4q}{(2\pi)^4} \frac{1}{q^2 - m^2 + i\epsilon} \frac{1}{(p-q)^2 - m^2 + i\epsilon}. \tag{5.27}$$

This integral is actually quite subtle. One can tell that it is dimensionless. We calculate this integral in the appendix, and it turns out to consist of a dimensionless constant plus a logarithm of a function of p^2. We will discuss this integral further in section 7.1, which discusses measurement, and it is further explored in section 7.3. For the time being we will leave $I(p)$ expressed simply as an integral.

There are two more diagrams with a similar topology to that of diagram (5.25), and they are obtained by attaching, respectively, legs 1 and 3 and legs 1 and 4 to the x vertex:

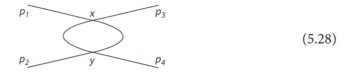

$$\tag{5.28}$$

and

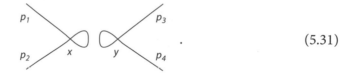

$$\tag{5.29}$$

which evaluate to

$$T_{fi}^{(2)}[\text{diagram (5.28)}] = -\frac{1}{2}\lambda^2 (2\pi)^4 \frac{\delta^{(4)}(p_1 + p_2 - p_3 - p_4)}{\sqrt{2\omega_1 \, 2\omega_2 \, 2\omega_3 \, 2\omega_4}} I(p_1 - p_3),$$

and

$$T_{fi}^{(2)}[\text{diagram (5.29)}] = -\frac{1}{2}\lambda^2 (2\pi)^4 \frac{\delta^{(4)}(p_1 + p_2 - p_3 - p_4)}{\sqrt{2\omega_1 \, 2\omega_2 \, 2\omega_3 \, 2\omega_4}} I(p_1 - p_4). \tag{5.30}$$

But there are more ways of connecting the parts in equation (5.24). For, instance, we might want to connect again two external legs to the x vertex and two to the y vertex, while connecting the remaining legs to the same vertices:

$$\tag{5.31}$$

It is easy to see that the evaluation of this diagram gives a result that is the same as that of equation (5.26) with one x and one y in the Feynman propagators exchanged (and a factor of $1/2$ from the reduction $\langle 0|T\left(\phi^2(x)\,\phi^2(y)\right)|0\rangle = \langle 0|T\left(\phi(x)\,\phi(x)\right)|0\rangle\,\langle 0|T\left(\phi(y)\,\phi(y)\right)|0\rangle$)

$$T_{fi}^{(2)}[\text{diagram (5.31)}] = -\frac{1}{2}\left(\frac{\lambda}{4!}\right)^2 \times 2 \times (4\times 3)^2$$

$$\times \int d^4x\,d^4y\,\frac{e^{-i(p_1+p_2)\cdot x}}{\sqrt{2\omega_{p_1}}\,\sqrt{2\omega_{p_2}}}\,\frac{e^{i(p_3+p_4)\cdot y}}{\sqrt{2\omega_{p_3}}\,\sqrt{2\omega_{p_4}}}$$

$$\times \int \frac{d^4q}{(2\pi)^4}\frac{e^{iq\cdot(x-x)}}{q^2-m^2+i\epsilon}\int \frac{d^4q'}{(2\pi)^4}\frac{e^{iq'\cdot(y-y)}}{q'^2-m^2+i\epsilon}$$

$$\propto \delta^{(4)}(p_1+p_2)\,\delta^{(4)}(p_3+p_4)\,, \qquad (5.32)$$

that vanishes. In fact, the result is proportional to $\delta(E_1+E_2)$, whose argument, the sum of the energies of the incoming particles, never vanishes.

One other possible topology is

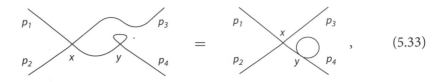

$$(5.33)$$

for which we get a factor of 2 to account for the equivalent diagram with $x \leftrightarrow y$; a factor $4\times 3\times 2$ by the various ways we can attach the three external legs to the vertex x; and corresponding factors $e^{-ip_2\cdot x}/\sqrt{2\omega_{p_2}}$, $e^{-ip_2\cdot x}/\sqrt{2\omega_{p_2}}$, and $e^{ip_3\cdot x}/\sqrt{2\omega_{p_3}}$. We also get a factor $e^{ip_4\cdot y}/\sqrt{2\omega_{p_4}}$ by attaching the p_4 leg to the y vertex. We are then left with $\langle 0|T\left(\phi(x)\,\phi^3(y)\right)|0\rangle = 3\,\langle 0|T\left(\phi(x)\,\phi(y)\right)|0\rangle\,\langle 0|T\left(\phi(y)\,\phi(y)\right)|0\rangle$. All in all, we finally obtain

$$T_{fi}^{(2)}[\text{Diagram (5.33)}] = -\frac{1}{2}\left(\frac{\lambda}{4!}\right)^2 \times 2 \times (4\times 3\times 2)\times 3$$

$$\times \int d^4x\,d^4y\,\frac{e^{-i(p_1+p_2-p_3)\cdot x}}{\sqrt{2\omega_{p_1}}\,\sqrt{2\omega_{p_2}}\,\sqrt{2\omega_{p_3}}}\,\frac{e^{ip_4\cdot y}}{\sqrt{2\omega_{p_4}}}$$

$$\times \int \frac{d^4q}{(2\pi)^4}\frac{e^{iq\cdot(x-y)}}{q^2-m^2+i\epsilon}\int \frac{d^4q'}{(2\pi)^4}\frac{e^{iq'\cdot(y-y)}}{q'^2-m^2+i\epsilon}$$

$$= -\frac{1}{2}\lambda^2\,(2\pi)^4\frac{\delta^{(4)}(p_1+p_2-p_3-p_4)}{\sqrt{2\omega_1\,2\omega_2\,2\omega_3\,2\omega_4}}\,\frac{1}{p_4^2-m^2+i\epsilon}$$

$$\int \frac{d^4q}{(2\pi)^4}\frac{1}{q^2-m^2+i\epsilon}\,. \qquad (5.34)$$

Then we will have equivalent diagrams with the y-based loop on the p_1, p_2, and p_3 legs. However, we will not be concerned about this class of diagrams, because, as we will see in section 5.5, the process of renormalization makes them effectively vanish when we describe the external states with the proper mass and normalization.

Finally, there are disconnected diagrams such as

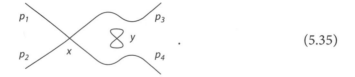

$$(5.35)$$

We will show in section 5.5 that these diagrams do not contribute to the matrix element, so we will ignore them from now on.

To sum up, only the diagrams (5.25), (5.28), and (5.29) give a nonvanishing contribution to our amplitude, so that our final result, including the first-order term, reads

$$-i\mathcal{M} = -i\lambda - \frac{1}{2}\lambda^2[I(p_1 + p_2) + I(p_1 - p_3) + I(p_1 - p_4)].$$ (5.36)

Let us now see why we can ignore the disconnected diagrams.

5.5 Getting rid of disconnected diagrams

Here is a welcome simplification: we will show that all the disconnected diagrams, such as diagram (5.35), drop out of the amplitudes, so we will not have to deal with them. Our starting point is the calculation of the *vacuum-to-vacuum transition amplitude*, whose absolute value squared is the probability that, if the system starts in its vacuum, it will also end in the same state. Using the general formula in equation (5.7), this reads

$$\langle 0|U_I(\infty, -\infty)|0\rangle = 1 + \infty + \frac{\infty}{\infty} + \ldots .$$ (5.37)

There will be additional types of diagrams appearing at order λ^2 and beyond. It should become clear how to draw these once we have completed the Feynman rules. A key point here is that the vacuum cannot decay into anything,[2] which implies that

$$|\langle 0|U_I(\infty, -\infty)|0\rangle|^2 = 1 \Rightarrow \langle 0|U_I(\infty, -\infty)|0\rangle = e^{-i\alpha}$$ (5.38)

for some real α. Because these diagrams without external legs are built out of Feynman propagators only, one can calculate that the phase really amounts to

$$\alpha = \frac{\lambda}{4} \int d^4x\, D_F^2(x, x)$$ (5.39)

---

[2]This is not always the case, because sometimes people use the term "vacuum" to describe some state that is metastable (i.e., stable classically, but not quantum mechanically), or even unstable but very long-lived. However, we will not consider this possibility here.

at first order in λ. Now, it is easy to see that disconnected diagrams correspond to amplitudes that are factorizable. For instance, diagram (5.35) is factorized into the product of two individual contributions

$$-i\left(\frac{\lambda}{4!}\right)^2 \int d^4x\,d^4y\,\langle p_3,\,p_4\,|\phi^4(x)|\,p_1,\,p_2\rangle\langle 0\,|\phi^4(y)|\,0\rangle = \underset{p_2 \quad p_4}{\overset{p_1 \quad p_3}{\times}}_x \times \,\underset{y}{8}.$$

$$(5.40)$$

Now, an incredible simplification emerges from the fact that the sum of all (connected and disconnected) diagrams can be factorized into the sum of connected diagrams times the sum of disconnected ones. Schematically,

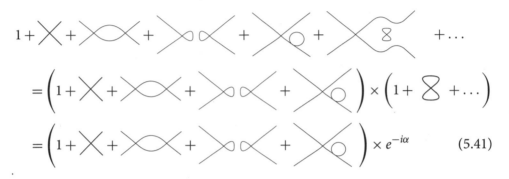

$$(5.41)$$

so that the disconnected diagrams factor out in a way that simply multiplies our matrix element by an irrelevant overall phase. For this reason, we can just go ahead and ignore those diagrams from now on.

5.6 The Feynman rules

From the discussion in section 5.4, it is natural to guess that there must be some protocol that allows us to associate each diagram to a formula giving the corresponding contribution to the amplitude. Indeed this is the case, and this protocol is given by the *Feynman rules*. Let us review them in the case of the $\lambda\phi^4$ theory which we have seen before.

1. For a process with N_i initial particles and N_f external particles, draw and label N_i incoming lines on the left and N_f outgoing lines on the right (time flows from left to right). Associate to each line a four-momentum p_i directed as in this diagram:

$$
\begin{array}{ccc}
p_1 & & p_{N_i+1} \\
p_2 & & p_{N_i+2} \\
\vdots & & \vdots \quad ; \\
p_{N_i} & & p_{N_i+N_f}
\end{array}
\qquad (5.42)
$$

2. for a calculation of the $O(\lambda^n)$ terms in $-i\mathcal{M}_{fi}$, draw n *vertices* with four legs:

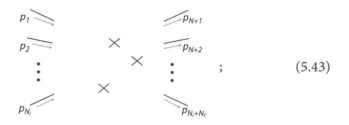

$$(5.43)$$

3. join all the external lines and all the vertices in all possible *fully connected* topologies (for instance, you should not include diagrams (5.31) and (5.35) and similar ones). Associate to each internal line a momentum q_i with an arbitrary direction:

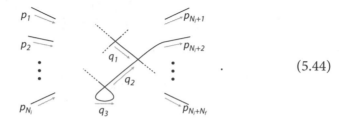

$$(5.44)$$

In simple processes, there might be only one diagram with the minimal number of vertices. Most commonly there are many diagrams that are possible. Each represents a contribution to the overall amplitude. We are now finished drawing, and we can start computing the value of each diagram. To do this:

4. Include a factor of $-i\lambda$ for each vertex;
5. for each scalar particle in an internal leg with momentum q_i, multiply by

$$\frac{i}{q^2 - m^2 + i\epsilon} \ ; \qquad (5.45)$$

6. impose momentum conservation at each vertex by multiplying by $(2\pi)^4 \delta^{(4)}(\sum_{i=1}^{4} k_i)$, where k_i is one of the momenta (be it an external momentum, p, or an internal one, q), including its sign, entering the vertex;
7. integrate over each internal momentum q_i by introducing a factor

$$\int \frac{d^4 q}{(2\pi)^4} \ ; \qquad (5.46)$$

8. if there are two identical scalars entering one loop, multiply by a factor of $1/2$. This is to account for the fact that for indistinguishable particles the integral

$$\int \frac{d^4q}{(2\pi)^4} \frac{i}{q^2 - m^2 + i\epsilon} \frac{i}{(q - p_1 - p_2)^2 - m^2 + i\epsilon} \qquad (5.47)$$

accounts for both diagrams

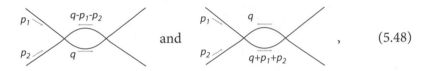

and   ,   (5.48)

which, however, are the same diagram.

9. The result you obtain gives $-i(2\pi)^4 \delta \left(\sum_{k=1}^{N_i} p_k - \sum_{k=1}^{N_f} p_{N_i + k} \right) \mathcal{M}_{fi}$.

10. Finally, the contribution of any diagram to the transition matrix T is given, as we have seen in section 4.5, by

$$T_{fi} = (2\pi)^4 \frac{\delta^{(4)} \left(\sum_{i=1}^{N_i} p_i - \sum_{i=1}^{N_f} p_{N_i + i} \right)}{\sqrt{2E_1 \dots 2E_{N_i + N_f}}} \mathcal{M}_{fi} . \qquad (5.49)$$

When there are several diagrams for a given process, we calculate each of them using these rules and add them together to get the overall matrix element.

The Feynman rules for all theories follow this pattern. One starts by drawing all possible diagrams at a given order in the perturbative expansion. The forms of these diagrams vary for different theories because the forms of the interactions are different. But we have already learned how to identify the basic vertices that follow from the interaction Lagrangian. For a given theory we take these vertices and start drawing. Then, because for each vertex there is a coupling constant, which we have identified by taking the basic matrix element, step 4 will involve adding a coupling constant associated with each vertex. Steps 5 through 10 are the same for all theories. If there are fermions or photons involved, there will be specific propagators for these particles. So the Feynman rules for other theories just involve minor variations on the first few steps, which are specific to that theory.

In the preceding description, we have tried to be precise in spelling out the steps in the Feynman rules. However, a professional field theorist would not feel the need to be so detailed. Because the feature that is unique to this theory is the vertex, a field theorist would generally just give the rule

$$= -i\lambda . \qquad (5.50)$$

The other steps (drawing all connected diagrams, inserting propagators for internal lines, integrating over all loop momenta, the symmetry factor) would be understood by other field theorists.

current, and we recall seeing these features in the treatment of the complex scalar field in section 3.9. To include the electromagnetic field we use the *gauge covariant derivative*

$$D_\mu \phi(x) = [\partial_\mu - iqA_\mu(x)]\phi(x), \tag{5.56}$$

where q is the particle's electric charge. The complex scalar had a $U(1)$ symmetry, which can here be turned into a $U(1)$ gauge symmetry,

$$\phi(x) \to \phi'(x) = e^{iq\chi(x)}\phi(x). \tag{5.57}$$

In contrast to our previous treatment, here we take the phase to have an arbitrary spacetime dependence in the real function $\chi(x)$. This would normally be a problem because derivatives of $\phi'(x)$ would end up involving $\partial_\mu \chi(x)$. The covariant derivative solves this problem, because if we simultaneously transform the electromagnetic field by a gauge transformation, as in equation (3.77), we have

$$\begin{aligned} D'_\mu \phi' &= \left[\partial_\mu - iqA'_\mu\right]\phi' \\ &= e^{iq\chi(x)}\left[\partial_\mu + iq(\partial_\mu \chi) - iq(A_\mu + \partial_\mu \chi)\right]\phi \\ &= e^{iq\chi(x)}D_\mu \phi. \end{aligned} \tag{5.58}$$

The factors of $\partial_\mu \chi(x)$ have been canceled and the derivative transforms in the same way as ϕ does, hence the name *covariant*. It is now easy to write a gauge-invariant Lagrangian,

$$\mathcal{L}_{\text{QED}} = -\frac{1}{4}F_{\mu\nu}F^{\mu\nu} + (D_\mu \phi)^*D^\mu \phi - m^2\phi^*\phi, \tag{5.59}$$

which is unchanged under the simultaneous transformations of equations (3.77) and (5.57). This is the Lagrangian of Scalar Quantum Electrodynamics.

The Feynman rules for the photon coupling to the scalar field follow immediately. We have the $\phi\phi\gamma$ coupling of

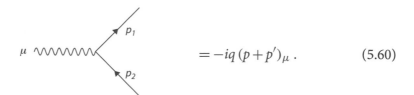

$$= -iq\,(p+p')_\mu. \tag{5.60}$$

There is also a vertex with two photons, with the rule

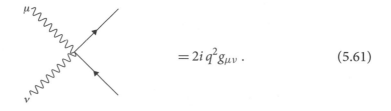

$$= 2i\,q^2 g_{\mu\nu}. \tag{5.61}$$

It is a bit more subtle to get the photon propagator. Because of the possibility of a gauge transformation, it can depend on the choice of gauge. The topic of gauge dependence is larger and more complex than we care to treat in this book. For us, we can just choose the Lorentz gauge, in which case Maxwell's equations become simply $\Box A_\nu = J_\nu$. Inverting this leads to the propagator

$$\mu \,\rlap{\,\wedge\!\wedge\!\wedge\!\wedge\!\wedge\!\wedge\!\wedge\!\wedge}{}\, \nu \qquad = \frac{-i g_{\mu\nu}}{q^2 + i\epsilon}\,, \qquad (5.62)$$

which completes the Feynman rules for scalar Quantum Electrodynamics.

While we are not going to use fermionic Quantum Electrodynamics in our development of Quantum Field Theory procedures, it is easy to give the Feynman rule for the theory. There is one vertex, given by

$$\mu \,\rlap{\,\wedge\!\wedge\!\wedge\!\wedge\!\wedge\!\wedge}{} \qquad\qquad = -iq\,\gamma_\mu\,, \qquad (5.63)$$

where γ_μ is a Dirac γ matrix. For fermions, there is no two-photon vertex like we saw for scalars in equation (5.61). As an example of applying this rule, we here write the matrix element for Rutherford scattering $e + p \to e + p$ with the same momentum conventions that we used in the scalar analog of this process in equation (5.16). The Feynman diagram is pictured the same way as that of the scalar analog, but the vertex is different. In addition we need to add on the spinors for the external particles. The result is

$$-i\mathcal{M} = (ie)\bar{u}_e(\mathbf{p}_3)\gamma^\mu u_e(\mathbf{p}_1)\,\frac{-i g_{\mu\nu}}{(p_1 - p_3)^2 + i\epsilon}\,(-ie)\bar{u}_p(\mathbf{p}_4)\gamma^\nu u_p(\mathbf{p}_2)\,. \qquad (5.64)$$

Here we have used $+e$ as the charge of the proton and have suppressed the spin labels on the Dirac spinors.

5.8 Relation with old-fashioned perturbation theory

While the Quantum Field Theory rules make sense on their own, they present a puzzle when compared to the rules for perturbation theory in ordinary quantum mechanics. In quantum mechanics we are taught to sum over intermediate states of different energies, that is, in second-order perturbation theory we have

$$\mathcal{M}_{fi} = \sum_I \frac{\langle f|V|I\rangle\langle I|V|i\rangle}{E - E_I}\,. \qquad (5.65)$$

In this sum, overall energy is conserved, $E_f = E_i \equiv E$, but energy is not conserved in the intermediate states. We rationalize this by saying that it is all right to give up

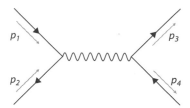

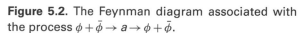

Figure 5.2. The Feynman diagram associated with the process $\phi + \bar{\phi} \rightarrow a \rightarrow \phi + \bar{\phi}$.

energy conservation for a short time, invoking the uncertainty principle $\Delta E \, \Delta t \geq \hbar$. These intermediate states are regular energy eigenstates, that is, they are on-shell. In contrast, the rules that we derived in section 5.6 have energy and momentum conserved at each vertex. That forces the particles that exist in the intermediate states to violate the $E^2 = \mathbf{p}^2 + m^2$ condition. That means the particles are off-shell. So which is it? Is energy conserved at all times or not? Are particles on-shell or off-shell? Amazingly, both descriptions are identical and are simultaneously valid.

Let us work out a specific example, using the amplitude associated with the diagram $\phi + \bar{\phi} \rightarrow a \rightarrow \phi + \bar{\phi}$, whose Feynman diagram is given in figure 5.2.

The Feynman rules that we have outlined in section 5.6 give the result

$$-i(2\pi)^4 \delta(p_1 + p_2 - p_3 - p_4) \, \mathcal{M}_{fi} = (-ig)^2 \int \frac{d^4 q}{(2\pi)^4} \frac{i}{q^2 - m_a^2}$$

$$\times (2\pi)^4 \delta^{(4)}(p_1 + p_2 - q) \, (2\pi)^4 \delta^{(4)}(q - p_3 + p_4) \,, \tag{5.66}$$

so that

$$\mathcal{M}_{fi} = \frac{g^2}{(p_1 + p_2)^2 - m_a^2} \,. \tag{5.67}$$

Energy conservation is respected at each vertex, so that the intermediate a particle carries energy and momentum $(E_1 + E_2, \mathbf{p}_1 + \mathbf{p}_2)$.

In contrast, in usual quantum mechanical perturbation theory, at leading order in g, there are two possible intermediate states. Because in this case

$$V = g \int d^3 x \, a(x) \, \phi^*(x) \, \phi(x) \,, \tag{5.68}$$

and the initial and final states are given explicitly as $|i\rangle = |\phi(p_1), \bar{\phi}(p_2)\rangle$ and $\langle f| = \langle \phi(p_3), \bar{\phi}(p_4)|$, then the intermediate states $|I\rangle$ for which $\langle I|V|i\rangle$ and $\langle f|V|I\rangle$ do not vanish are of the form

$$|I\rangle_a = |a\rangle \qquad |I\rangle_b = |a, \phi, \phi, \bar{\phi}, \bar{\phi}\rangle \,, \tag{5.69}$$

so that $|I\rangle_a$ is a single a state, and $|I\rangle_b$ is a state containing five particles. One can represent the contributions proportional to $\langle f|V|I\rangle_a \,_a\langle I|V|i\rangle$ and $\langle f|V|I\rangle_b \,_b\langle I|V|i\rangle$, respectively, as

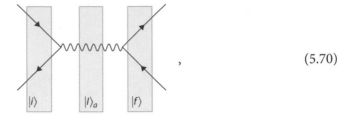

, (5.70)

and

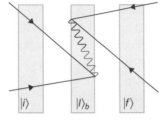

, (5.71)

where we have shaded the portions of diagram that correspond to initial, intermediate, and final states.

Thanks to the three-dimensional integral in the interaction term in equation (5.68), momentum is conserved at every vertex in the calculation (even though energy is not). This means that the kinematics is the following. For the initial state and the final state the energy is $E = E_1 + E_2 = E_3 + E_4$. Given momentum conservation, the energy of the on-shell intermediate state in diagram (5.70) is

$$E_a = \sqrt{(\mathbf{p}_1 + \mathbf{p}_2)^2 + m_a^2},$$ (5.72)

whereas for diagram (5.71) all the ϕ and $\bar{\phi}$ particles are on-shell so that

$$E_b = E_1 + E_2 + E_3 + E_4 + \sqrt{(-\mathbf{p}_1 - \mathbf{p}_2)^2 + m_a^2}.$$ (5.73)

Now, let us calculate the transition amplitude. The matrix element for the entry to the intermediate state in diagram (5.70) is found to be

$$\langle a(p_a)|V|\phi(p_1),\, \bar{\phi}(p_2)\rangle = \int d^3x\, g\, \frac{e^{i(\mathbf{p}_1 + \mathbf{p}_2 - \mathbf{p}_a)\cdot\mathbf{x}}}{\sqrt{2E_1\, 2E_2\, 2E_a}}\, e^{i(E_a - E_1 - E_2)t}$$

$$= g\,(2\pi)^3 \frac{\delta^{(3)}(\mathbf{p}_1 + \mathbf{p}_2 - \mathbf{p}_a)}{\sqrt{2E_1\, 2E_2\, 2E_a}}\, e^{i(E_a - E_1 - E_2)t}.$$ (5.74)

As usual we will discard the $1/\sqrt{2E}$ factors for the external state particles, but we keep it for the internal particle. The matrix element to transition to the final state is the complex conjugate of this amplitude, but with momenta $\mathbf{p}_3$ and $\mathbf{p}_4$.

For diagram (5.71), the calculation is complicated by the larger number of momenta that appear in the matrix element $\langle a(p_a),\, \phi(p_b),\, \phi(p_c),\, \bar{\phi}(p_d),\, \bar{\phi}(p_e)|V|$

$\phi(p_1), \bar{\phi}(p_2)\rangle$, giving terms proportional to $\delta^{(3)}(\mathbf{p}_a + \mathbf{p}_b + \mathbf{p}_d)\, \delta^{(3)}(\mathbf{p}_c - \mathbf{p}_1)$ $\delta^{(3)}(\mathbf{p}_e - \mathbf{p}_2)e^{i(E_a+E_b+E_d)t}$, as well as permutations. In the end, however, if we require that the final momenta differ from the initial ones, then only one combination survives. In both cases we see that, as announced, the three-momentum is conserved at all steps. Integrating over all possible momenta in the intermediate state, and noticing that time dependence disappears when we impose energy conservation $E_1 + E_2 = E_3 + E_4$, we are left with

$$(2\pi)^3\delta^{(3)}(\mathbf{p}_1 + \mathbf{p}_2 - \mathbf{p}_3 - \mathbf{p}_4)\, \mathcal{M}_{fi} = \int \frac{d^3 p_a}{(2\pi)^3} \frac{1}{2E_a}$$

$$\times \left[\frac{g\,(2\pi)^3\,\delta^{(3)}(\mathbf{p}_1 + \mathbf{p}_2 - \mathbf{p}_a)\,g\,(2\pi)^3\,\delta^{(3)}(\mathbf{p}_3 + \mathbf{p}_4 - \mathbf{p}_a)}{E - E_a} \right.$$

$$\left. + \frac{g\,(2\pi)^3\,\delta^{(3)}(\mathbf{p}_1 + \mathbf{p}_2 + \mathbf{p}_a)\,g\,(2\pi)^3\,\delta^{(3)}(\mathbf{p}_3 + \mathbf{p}_4 + \mathbf{p}_a)}{E - E_b} \right]. \quad (5.75)$$

The matrix element is then

$$\mathcal{M}_{fi} = \frac{1}{2E_{12}} \left[\frac{g^2}{E_1 + E_2 - E_{12}} + \frac{g^2}{E_1 + E_2 - (E_1 + E_2 + E_3 + E_4 + E_{12})} \right], \quad (5.76)$$

where $E_{12} = \sqrt{(\mathbf{p}_1 + \mathbf{p}_2)^2 + m_a^2}$. Using $E_1 + E_2 = E_3 + E_4$ and combining energy denominators, we have

$$\mathcal{M}_{fi} = \frac{g^2}{(E_1 + E_2)^2 - E_{12}^2} = \frac{g^2}{(E_1 + E_2)^2 - (\mathbf{p}_1 + \mathbf{p}_2)^2 - m_a^2} = \frac{g^2}{(p_1 + p_2)^2 - m_a^2}, \quad (5.77)$$

which is identical to the Feynman diagram approach! The "conserve energy at every vertex with off-shell intermediate states" version and the "violate energy conservation for a short time in the intermediate state but use on-shell states" version are mathematically identical. This is good food for thought.

The origin of this remarkable identity is the connection between the two forms that we derived for the propagator, one with time ordering, equation (4.41), and one in covariant form, equation (4.44). In equation (4.41), the energies satisfy the on-shell condition, and only the three-momentum is integrated over. If the intermediate state energy must be on-shell, it is not possible to conserve energy at each vertex. In fact, if we use the time-ordered form of equation (4.41) in our derivation of the transition amplitude, we would rederive the quantum mechanical perturbation theory result that we just calculated. In the covariant case, the energy variable ranges over all values. Using this form, the integrations over spacetime enforce energy and momentum conservation at each vertex. It is this form that is used when the energy is conserved at each vertex. However, using a contour integration, we showed that both of these forms are mathematically equivalent. Both expressions have identical physics, even though it does not seem readily apparent.

The field theory method is much more efficient. In this example, the single Feynman diagram contained the physics of two time-ordered diagrams. In more complicated processes, the simplifications would be greater. From this example, we also learn that we should not think of Fenyman diagrams as direct pictures of what is happening in spacetime. They express the connectedness of the physical process but do not indicate what happens earlier or later in time.

Chapter summary: In this chapter, we have seen how perturbation theory arises in Quantum Field Theory. The interactions cause the initial state to evolve, with transitions happening in ways determined by the interaction terms in the Lagrangian. It was quite a bit of work to sort through all the possibilities, but the end result is a simple set of Feynman rules. You do not have to go through all the hard work each time. With practice, you can write down the amplitude in a few seconds. The (slightly simplified) procedure is that you draw the connected diagrams, putting a known coupling at each vertex and a propagator for each internal line. For simple amplitudes (tree diagrams), you are finished after these steps. For loop diagrams, you still need to evaluate the loop over the internal momentum. But at least you have identified the desired matrix element.

CHAPTER 6

Calculating

Once we have the Feynman rules, we can, in principle, calculate all the transition amplitudes we want. This chapter is a sample of some of the physical predictions we can get from Quantum Field Theory.

6.1 Decay rates and cross sections

For a start, we can turn the amplitudes into observables, relating the amplitudes to decay rates and cross sections.

For ready reference, here are the results for a decay $A \to B + C$ in our conventions

$$
\Gamma(A \to B + C) = \frac{1}{2M_A} \frac{1}{S} \int \frac{d^3 p_B}{(2\pi)^3} \frac{1}{2E_B} \frac{d^3 p_C}{(2\pi)^3} \frac{1}{2E_C}
$$
$$
\times |\mathcal{M}|^2 \, (2\pi)^4 \delta^{(4)}(p_A - p_B - p_C) \tag{6.1}
$$

and also for a scattering cross section $A + B \to C + D$

$$
\sigma(A + B \to C + D) = \frac{1}{2E_A} \frac{1}{2E_B} \frac{1}{|\mathbf{v}_A - \mathbf{v}_B|} \frac{1}{S}
$$
$$
\times \int \frac{d^3 p_C}{(2\pi)^3} \frac{1}{2E_C} \frac{d^3 p_D}{(2\pi)^3} \frac{1}{2E_D} |\mathcal{M}|^2 \, (2\pi)^4 \delta^{(4)}(p_A + p_B - p_C - p_D) \tag{6.2}
$$

In these formulas, $\mathcal{M}$ is the matrix element with the normalization conventions used in this book. In addition, S is a symmetry factor, which is normally equal to 1, except for when there are two identical particles in the final state, in which case it is equal to 2. Let us now see where these relations come from.

When dealing with rates for processes in quantum mechanics we turn to Fermi's Golden Rule. Despite their more formidable appearance, equations (6.1) and (6.2) are really just elaborations of the Golden Rule. We will first quickly review the quantum mechanical formulation and then extend it to the quantum field–theoretical setup.

Rigorous derivations of Fermi's Golden Rule require care with the limiting procedures to identify the time dependence. What follows is a quick and less careful path to obtain the same result. If we assume a potential interaction of the form $V(t) = V_I \, e^{i\omega t}$, then the transition amplitude is given by

$$\text{Amp}_{fi} = \int dt \, \langle f | V_I(t) | i \rangle \, e^{i(E_f - E_i)t}$$

$$= 2\pi \, \langle f | V_I | i \rangle \, \delta(E_i - E_f - \omega) \,. \qquad (6.3)$$

To get a probability we square it:

$$\text{Prob}(i \rightarrow f) = |\text{Amp}_{fi}|^2 = |\langle f | V_I | i \rangle|^2 \, [2\pi \, \delta(E_i - E_f - \omega)]^2 \,. \qquad (6.4)$$

The product of two delta functions can be interpreted by treating one of them as

$$2\pi \, \delta(0) = \int dt \, e^{-i(0)t} = \int dt = T \,, \qquad (6.5)$$

where we have enforced $E_i - E_f - \omega = 0$ from the other delta function, and where T is the entire duration of the measurement. This yields

$$\text{Rate}(i \rightarrow f) = \frac{\text{Prob}(i \rightarrow f)}{T} = 2\pi \, |\langle f | V_I | i \rangle|^2 \, \delta(E_i - E_f - \omega) \,, \qquad (6.6)$$

which is Fermi's Golden Rule in its most basic form. For time-independent interaction terms, such as those that we will use, we can just send $\omega \rightarrow 0$. While the math has been quick and dirty, the result is correct. It even makes intuitive sense, with the amplitude-squared factor and energy conservation.

Equation (6.6) is fine if we want to study the transition from some initial state to an isolated final state. But what if there are many final states, possibly a continuum, that are accessible to the system? In that case, we have to sum on all those final states:

$$\text{Rate}(i \rightarrow f) = 2\pi \sum_j |\langle f_j | V_I | i \rangle|^2 \, \delta(E_i - E_{f_j}) \,. \qquad (6.7)$$

In terms of the *density of states* $\rho_f(E)$, defined in such a way that there are $\rho_f(E) \, dE$ final states with energy between E and $E + dE$, we replace

$$\sum_j \rightarrow \int \rho_f(E_f) \, dE_f \qquad (6.8)$$

in equation (6.7), obtaining at last

$$\text{Rate}(i \to f) = 2\pi \, |\langle f|V_I|i\rangle|^2 \, \rho_f(E_f) \,, \tag{6.9}$$

where we used $E_f = E_i$. We are now left with the task of computing $\rho_f(E_f)$. If the system is in a large but finite volume V, with periodic boundary conditions for the wavefunctions, the number of states with momenta between $\mathbf{p}_f$ and $\mathbf{p}_f + d\mathbf{p}_f$, for each particle, is

$$dN_f = V \frac{d^3 p_f}{(2\pi)^3} \,. \tag{6.10}$$

Each of the final momentum states carries a normalization factor of $1/\sqrt{V}$, so that the volume factors drop out in the final result. Including the sum over final states, Fermi's Golden Rule becomes

$$\text{Rate} = \int \frac{d^3 p_f}{(2\pi)^3} \, |\langle f|V_I|i\rangle|^2 \, (2\pi)\delta(E_i - E_f) \,, \tag{6.11}$$

which is the form used in most nonrelativistic quantum-mechanical applications. In this form, the density of final states becomes

$$\rho_f(E_f) = \frac{dN_f}{dE_f} = \frac{p_f^2 \, dp_f \, d\Omega}{(2\pi)^3 \, dE_f} = \frac{p_f \, E_f}{(2\pi)^3} \, d\Omega \,, \tag{6.12}$$

where we have used $p \, dp = E \, dE$, that originates from $p^2 + m^2 = E^2$, and where $d\Omega$ is the element of solid angle.

However, even this form of the rule is incomplete. For example, when applied to the atomic decay $A^* \to A + \gamma$, the final state factor sums over all photon final states but not final atomic states. Why this asymmetry? In this case, the final atom is assumed to be at rest in the usual quantum-mechanical treatment. For a general decay $A \to B + C$, we should treat both particles in the final state on equal footing. The symmetric density of final states factor sums over both final states, which is consistent with momentum conservation,

$$\text{Rate}(A \to B + C) = \int \frac{d^3 p_B}{(2\pi)^3} \frac{d^3 p_C}{(2\pi)^3} \, |\langle f|V_I|i\rangle|^2$$
$$\times (2\pi)^3 \delta^{(3)} \left(\mathbf{p}_A - \mathbf{p}_B - \mathbf{p}_C \right) (2\pi) \, \delta \left(E_A - E_B - E_C \right) \,, \tag{6.13}$$

which reduces to equation (6.11) when the amplitude does not depend on the momentum of one of the final state particles.

Finally, we recall that in forming the Feynman rules we have stripped out the various factors of $1/\sqrt{2E}$ associated with external states, via

$$\langle B \, C|V_I|A\rangle = \frac{1}{\sqrt{2E_A}} \frac{1}{\sqrt{2E_B}} \frac{1}{\sqrt{2E_C}} \, \mathcal{M} \,. \tag{6.14}$$

Incorporating these factors and noting that the decay rate of a particle is defined as the rate of a particle at rest, we arrive at our final result, equation (6.1). We have chosen this pathway to equation (6.1) to emphasize that the result is just Fermi's Golden Rule presented as a more general form.

There is a small complication in the somewhat rare case that two final state particles are identical. Because the particles are then indistinguishable, the phase space factor

$$\int \frac{d^3 p_B}{(2\pi)^3} \frac{d^3 p_C}{(2\pi)^3} \tag{6.15}$$

overcounts the final states by a factor of 2. For every possible configuration where p_B and p_C occur in the integral, there is an identical configuration with the values of p_B and p_C exchanged. To account for this overcounting, we need to divide the integral by a factor of 2. This is accounted for by a symmetry factor S where $S = 1$ for nonidentical particles and $S = 2$ for two identical particles. If we were to generalize the formula to n particles in the final state, the symmetry factor would be $n!$ if they were all identical.

The factors of

$$\int \frac{d^3 p}{(2\pi)^3} \frac{1}{2E} \tag{6.16}$$

are sometimes referred to as *Lorentz-invariant phase space*. While they do not look obviously Lorentz invariant in this form, they can be rewritten as

$$\int \frac{d^4 p}{(2\pi)^4} (2\pi)\, \delta(p_\mu p^\mu - m^2)\, \Theta(p_0), \tag{6.17}$$

which is obviously Lorentz invariant once we recognize that boosts and rotations cannot change the sign of the energy variable. If there are more than two particles in the final state, you just include additional factors of the phase space integration, equation (6.16), for each particle into the formula for the decay rate (or the cross section) and include that particle's momentum in the energy-momentum conserving delta function.

The cross section formula is obtained with a very similar calculation. The definition of a cross section is Rate/Flux, and we can again calculate this using Fermi's Golden Rule. In the nonrelativistic quantum-mechanical case of a particle scattering off a fixed target, temporarily using box normalization, we have the single particle flux being v_i/V (where v_i is the initial velocity of the particle) as usual, and

$$d\sigma = \frac{\text{Rate}}{\text{Flux}} = \frac{1}{v_i/V} \times V \int \frac{d^3 p_f}{(2\pi)^3} |\langle f | V_I | i \rangle|^2 (2\pi)\, \delta(E_f - E_i)$$

$$= \frac{1}{p_i/(mV)} \times V \int \frac{d^3 p_f}{(2\pi)^3} \left| \int d^3 r \frac{e^{-i\mathbf{p}_f \cdot \mathbf{r}}}{\sqrt{V}} V_I(\mathbf{r}) \frac{e^{i\mathbf{p}_i \cdot \mathbf{r}}}{\sqrt{V}} \right|^2 (2\pi)\, \delta(E_f - E_i). \tag{6.18}$$

Again the volume factors drop out. This is actually a very familiar result. Using the density of final states factor from equation (6.12) in the nonrelativistic limit $E = m$

and $p_f = p_i$ as appropriate for elastic scattering, we obtain

$$\frac{d\sigma}{d\Omega} = \left| \frac{m}{2\pi} \int d^3r \, e^{-i(\mathbf{p}_f - \mathbf{p}_i) \cdot \mathbf{r}} \, V_I(\mathbf{r}) \right|^2 , \tag{6.19}$$

which is the standard formula from quantum mechanics.

To generalize equation (6.18) to a scattering process in Quantum Field Theory, we first note that the quantum-mechanical example did not include the recoil of the target, and we again sum over both final state particles, as in equation (6.13). Moreover, the velocity factor $|v_i|$ in the flux needs to be generalized to the relative velocity $|\mathbf{v}_A - \mathbf{v}_B|$. Finally we extract all of the normalization factors $1/\sqrt{2E_i}$ in defining the matrix element. Accomplishing these leads to the result quoted at the start of this section, equation (6.2).

6.2 Some examples

There are some calculational methods that are used in applying these formulas. These can be seen in explicit examples.

6.2.1 DECAY RATE

Let us calculate a simple example of a decay rate using our full relativistic formalism. Consider a heavy spinless particle σ with mass M_σ, which can decay into two lighter spinless particles χ with mass $m_\chi < M_\sigma/2$. This could be described by an interaction term

$$\mathcal{L}_I = -\frac{g}{2} \sigma \chi^2 v, \tag{6.20}$$

which leads to the matrix element

$$\mathcal{M} = g . \tag{6.21}$$

Using your skills with dimensional analysis, you will note that the coupling constant g carries the dimension of a mass. To complete the calculation of the decay rate, all that is left to do is to calculate the phase space factors because

$$\Gamma(\sigma \to 2\chi) = \frac{1}{2M_\sigma} \frac{1}{2} \int \frac{d^3 p_1}{(2\pi)^3} \frac{1}{2E_1} \frac{d^3 p_2}{(2\pi)^3} \frac{1}{2E_2} \, g^2 \, (2\pi)^4 \, \delta^{(4)}(p_\sigma - p_1 - p_2) . \tag{6.22}$$

Here we have also included the symmetry factor $S = 2$ because the two final state particles are identical. One of the momentum integrals can be trivially done using the momentum delta function to yield

$$\Gamma(\sigma \to 2\chi) = \frac{g^2}{2M_\sigma} \frac{1}{2} \int \frac{d^3 p_1}{(2\pi)^3} \frac{1}{2E_1} \frac{1}{2E_2} \, (2\pi) \, \delta (M_\sigma - E_1 - E_2) . \tag{6.23}$$

Because we are in the rest frame of σ, the momentum conservation constraint reads $\mathbf{p}_2 = -\mathbf{p}_1$, so that $E_2 = E_1 = M_\sigma/2$. Using $E^2 = \mathbf{p}^2 + m^2$, we have $p_1\, dp_1 = E_1\, dE_1$. This allows the final integrals to be done as

$$\int \frac{d\Omega\, p_1\, E_1\, dE_1}{(2\pi)^3} \frac{1}{(2E_1)^2} (2\pi)\, \delta(M_\sigma - 2E_1) = \frac{p_1\, E_1}{2\pi\, M_\sigma^2}. \qquad (6.24)$$

Finally, using $p_1 = \sqrt{E_1^2 - m_\chi^2}$ with $E_1 = M_\sigma/2$, we end up with the total decay rate

$$\Gamma = \frac{g^2}{32\pi\, M_\sigma} \sqrt{1 - 4\frac{m_\chi^2}{M_\sigma^2}}. \qquad (6.25)$$

Dimensional analysis confirms that this has the appropriate dimension for a decay rate.

6.2.2 CROSS SECTION

Next, let us calculate the basic cross section in $\lambda\phi^4$ theory. This is most easily done in the center of mass frame, where $2E_A = 2E_B = 2E_C = 2E_D = E_{cm} = \sqrt{s}$. The matrix element is just the constant λ. There is the symmetry factor $S = 2$ because the two final state particles are identical. The momentum integrations are very similar to the decay rate in section 6.2.1. We have

$$d\sigma = \frac{1}{2E_A 2E_B |\mathbf{v}_A - \mathbf{v}_B|} \frac{1}{2} \int \frac{d^3 p_C}{(2\pi)^3 2E_C} \frac{d^3 p_D}{(2\pi)^3 2E_D} \lambda^2 (2\pi)^4 \delta^4(p_A + p_B - p_C - p_D)$$

$$= \frac{1}{\sqrt{s}\,\sqrt{s}} \frac{1}{2\, p/E} \frac{1}{2} \times \frac{p\, E}{2} \frac{\lambda^2}{(2\pi)^2} d\Omega, \qquad (6.26)$$

which yields

$$\frac{d\sigma}{d\Omega} = \frac{\lambda^2}{32\pi^2} \frac{1}{s}. \qquad (6.27)$$

6.2.3 COULOMB SCATTERING IN SCALAR QUANTUM ELECTRODYNAMICS

Let us take a look at how one can calculate nonrelativistic Coulomb scattering using the full framework of Quantum Field Theory. Because we are for the most part avoiding using Dirac fermions in examples, we will use complex scalar fields. The electron will be mimicked by a light complex scalar ϕ_e with mass m_e and charge $-e$, while the heavy complex scalar ϕ_p with mass M_p and charge $+e$ will play the role of the proton. We gave the Feynman rules in section 5.7. The scattering process at tree level is given in figure 6.1, and the corresponding matrix element is

$$-i\,\mathcal{M} = (-ie)\,(p_1 + p_3)^\mu \frac{-i g_{\mu\nu}}{q^2} (+ie)\,(p_2 + p_4)^\nu, \qquad (6.28)$$

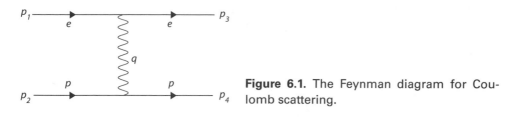

Figure 6.1. The Feynman diagram for Coulomb scattering.

with $q^2 = (p_1 - p_3)^2 = (p_2 - p_4)^2$. Because we will soon take the proton's mass to be large and thereby we will neglect recoil, it is simplest to calculate this cross section in the "lab" frame where the initial heavy particle is at rest, $E_2 = M_p$. In this case the velocity factor refers to the incoming light particle with $|\mathbf{v}_1 - \mathbf{v}_2| = p_1/E_1$, and we have

$$
\sigma = \frac{1}{2E_1 \, 2M_p} \frac{1}{p_1/E_1} \int \frac{d^3 p_3}{(2\pi)^3 2E_3} \frac{d^3 p_4}{(2\pi)^3 2E_4} |\mathcal{M}|^2 \, (2\pi)^4 \delta^{(4)}(p_1 + p_2 - p_3 - p_4)
$$

$$
= \frac{1}{2p_1 \, 2M_p} \int \frac{p_3 \, E_3 \, dE_3 \, d\Omega}{(2\pi)^3 \, 2E_3 \, 2E_4} |\mathcal{M}|^2 \, (2\pi) \, \delta(E_1 + M_p - E_3 - E_4) . \tag{6.29}
$$

While the integral over E_3 can be readily done in general, it saves some algebra at this point in the calculation to work in the heavy mass limit, $E_4 \sim M_p$. The energy delta function then becomes just $\delta(E_1 - E_3)$, which also implies $p_3 = p_1$. Using these relations, we obtain

$$
\frac{d\sigma}{d\Omega} = \frac{|\mathcal{M}|^2}{64\pi^2 \, M_p^2} . \tag{6.30}
$$

Kinematics also gives $(p_1 + p_3)^\mu (p_2 + p_4)_\mu = 4 E_1 M_p$ and $q^\mu q_\mu = -\mathbf{q}^2 = -(\mathbf{p}_1 \cdot \mathbf{p}_3)^2$, so that we end up with

$$
\frac{d\sigma}{d\Omega} = \left(\frac{e^2}{4\pi}\right)^2 \frac{4 E_1^2}{\mathbf{q}^4} \rightarrow \frac{4\alpha^2 \, m_e^2}{\mathbf{q}^4} , \tag{6.31}
$$

where in the final form we have taken the nonrelativistic limit for the ϕ_e field. This reproduces the usual nonrelativistic Coulomb scattering result. The calculation in Quantum Field Theory also gives the full relativistic result, so it is clearly more powerful than the standard quantum-mechanical calculation.

6.2.4 COULOMB POTENTIAL

The amplitude given in equation (6.28) can be related to the basic Coulomb potential governing nonrelativistic particles, which is the foundation for the quantum physics of atoms. To show this, we take the nonrelativistic limit of equation (6.28), in which case the matrix element is dominated by the $\mu = \nu = 0$ component of the four-momenta. We also need to reinstate the factors of $1/\sqrt{2E_i}$ for the external particles that we temporarily extracted when deriving the

Feynman rules. The resulting matrix element is

$$
\begin{aligned}
-i\tilde{\mathcal{M}} &\equiv \frac{-i\mathcal{M}}{\sqrt{2E_1}\sqrt{2E_2}\sqrt{2E_3}\sqrt{2E_4}} \\
&= \frac{1}{\sqrt{2E_1}\sqrt{2E_2}\sqrt{2E_3}\sqrt{2E_4}} \times e^2 \, (E_1 + E_3) \, \frac{i}{q^2} \, (E_2 + E_4) \simeq -i\frac{e^2}{\mathbf{q}^2} \, ,
\end{aligned}
\qquad (6.32)
$$

where the last equality follows from noting that in the nonrelativistic limit $E_1 \simeq E_3$ and $E_2 \simeq E_4$, such that $q^2 \simeq -\mathbf{q}^2$. The Coulomb potential is then just the Fourier transform of this amplitude

$$
V(r) = \int \frac{d^3 q}{(2\pi)^3} e^{i\mathbf{q}\cdot\mathbf{x}} \, \tilde{\mathcal{M}} = -\frac{e^2}{4\pi r} \, ,
\qquad (6.33)
$$

and you can convince yourself of this by inverting equation (6.19), that is used in nonrelativistic quantum mechanics to extract the amplitude from the potential.

Clearly, this procedure can be extended to other Lagrangians in Quantum Field Theory to obtain the corresponding nonrelativistic potential.

6.3 Symmetry breaking

Contributions from gradient terms to the energy of fields should come with a positive sign to keep the energy bounded from below. As a consequence, we expect that the energy is lowest when the fields are uniform. However, in several cases, there are qualitative differences between systems where the lowest energy state lies at vanishing values of the field and systems where the field does not vanish in vacuum. Ginzburg-Landau theories of magnetism or superconductivity provide a mechanism for the classical ground state of a field to have nonzero value. For example if the field $\mathbf{m}(\mathbf{x})$ denotes the local magnetization density, the Free Energy can be expanded as

$$
F \sim a \, |\mathbf{m}|^2 + b \, |\mathbf{m}|^4 + c \, |\nabla \mathbf{m}|^2 + \dots \, ,
\qquad (6.34)
$$

with $c > 0$. When the expansion is truncated at this order, the coefficient b should be positive so that the energy is positive for large magnetization. But if the coefficient a is negative, the minimum of the Free Energy occurs for a constant nonzero value of the magnetization density

$$
|\mathbf{m}(\mathbf{x})|^2 = -\frac{a}{2b} \, .
\qquad (6.35)
$$

Although the Free Energy is rotationally invariant, the implementation of this condition will force $\mathbf{m}$ to pick out a particular direction in the vacuum. This means that the rotational symmetry of the theory is broken in the vacuum. This simple model of *symmetry breaking* turns out to have very important realizations as full quantum field theories.

We can start from a symmetric theory. For example, let us take N real scalar fields ϕ^a with $a = 0, \ldots, N - 1$, with the labeling chosen for later convenience. In a symmetric theory, these fields would all enter the Lagrangian in the same way

$$\mathcal{L} = \frac{1}{2} \partial_\mu \phi^a \partial^\mu \phi^a - V\left(\phi^a \phi^a\right), \tag{6.36}$$

with, for instance,

$$V\left(\phi^a \phi^a\right) = \frac{m^2}{2} \phi^a \phi^a + \frac{\lambda}{4} \left(\phi^a \phi^a\right)^2, \tag{6.37}$$

where we have used the convention that a repeated index is understood to be summed upon. This model encompasses several possibilities. If $N = 2$, we recover the complex scalar field

$$\phi = \frac{1}{\sqrt{2}} \left(\phi_R + i\phi_I\right) \quad \text{and} \quad V(\phi) = m^2 |\phi|^2 + \lambda |\phi|^4. \tag{6.38}$$

If $N = 3$, this could be describing a vector field, such as the magnetization. The general N case is referred to as the $O(N)$ model.

When $m^2 > 0$, this is a normal symmetric theory. However in the case

$$m^2 = -\mu^2 < 0, \tag{6.39}$$

this will have a classical ground state with a nonzero magnitude, like the Ginzburg-Landau model does. Because this choice appears to correspond to a negative mass-squared, you could initially be puzzled about the content of this theory. However, we will see that after identifying the ground state, all masses will be real. With the kinetic energy being positive, the ground state will involve values of ϕ^a that are constant in space and time. The lowest energy states of these would minimize the field potential $V(\sum \phi^a \phi^a)$, which amounts to the condition

$$\sum_a \phi^a \phi^a = \frac{\mu^2}{\lambda} \equiv v^2. \tag{6.40}$$

While this condition itself exhibits the symmetry, there is no single field configuration satisfying it that is symmetric.

The real fun comes when we implement the ground state condition. In the general case, let us choose the 0-th field to have the nonzero value, remembering that all choices are equivalent by the initial symmetry,

$$\phi^0 \big|_{\text{ground state}} = v. \tag{6.41}$$

But this field will not only have the ground state value, but it also may have excitations, so we define the full field

$$\phi^0(x) = v + \sigma(x), \tag{6.42}$$

where in this new notation $\sigma(x)$ is the dynamical field. Rewriting the Lagrangian with this new field is a matter of simple algebra, and we find

$$\mathcal{L} = \frac{1}{2}\,\partial_\mu \phi^i \partial^\mu \phi^i + \frac{1}{2}\,[\partial_\mu \sigma \partial^\mu \sigma - 2\,\mu^2 \sigma^2] - \tilde{V}(\sigma, \phi^i \phi^i)\,, \qquad (6.43)$$

where now the index i goes over $N-1$ values, $i = 1, \ldots N-1$, and

$$\tilde{V}(\sigma, \phi^i \phi^i) = \lambda\,v\,\sigma(\sigma^2 + \phi^i \phi^i) + \frac{\lambda}{4}(\sigma^2 + \phi^i \phi^i)^2\,. \qquad (6.44)$$

Reading the content of this Lagrangian, we conclude that the $N-1$ fields ϕ^i are massless; the field σ is massive with mass $\sqrt{2}\,\mu$; and we see a new cubic interaction that was not there originally.

The fact that the σ now has a positive mass-squared is gratifying, because our initial starting point did not have this property. However the real news is the mass-lessness of the remaining fields. This is a consequence of the original symmetry. To understand the connection to the symmetry, we recall that the ground state expectation value in equation (6.41) could have been chosen in any direction in the space spanned by the ϕ^a coordinates. This implies that, if we were to shift the direction of the ground state (say, by an angle θ) from being entirely in the 0-th direction to having a small component in a different direction, say the j-th, there would be no change in the energy of the system. This requirement then forbids a $-m^2(\phi^j)^2$ term in the Lagrangian, as that term would raise the energy. This is the content of *Goldstone's theorem*—when the symmetry of the ground state does not reflect the symmetry of the Lagrangian, there are massless fields (often referred to as Goldstone bosons) in the physical spectrum.

The phenomenon described in this section is referred to as *spontaneous symmetry breaking*: a symmetry of the Lagrangian is not manifest in the ground state of the theory. The name is a bit of a misnomer, as the symmetry is still present although it is hidden. To demonstrate the hidden nature of the symmetry, let us take the notationally simplest case of a complex scalar field, as in equation (6.38). The symmetry of the theory is simply a phase change

$$\phi \rightarrow \phi' = e^{i\theta}\phi \qquad (6.45)$$

that corresponds to a rotation of the real and imaginary parts of the field. Choosing the ground state in the direction of the real part of the field and expanding

$$\phi = \frac{1}{\sqrt{2}}\,(v + \sigma + i\phi_I) \qquad (6.46)$$

leads to the same form as shown in equations (6.43) and (6.44) but with just a single Goldstone field $\phi^i = \phi_I$. In this form there is little visible remnant of the symmetry.

However, we could have used a different parameterization of the fields, by naming

$$\phi = \frac{1}{\sqrt{2}} (v + \bar{\sigma}) \, e^{i\chi/v} \tag{6.47}$$

with new fields $\bar{\sigma}(x)$ and $\chi(x)$. With these fields the original symmetry of equation (6.45) is simply

$$\chi(x) \to \chi'(x) = \chi(x) + v\theta \,, \tag{6.48}$$

which is just the shift of the χ field by a constant. In terms of these fields, the Lagrangian reads

$$\mathcal{L} = \frac{1}{2} \left[\partial_\mu \bar{\sigma} \partial^\mu \bar{\sigma} - m_\sigma^2 \bar{\sigma}^2 \right] + \frac{1}{2} \left(\frac{v + \bar{\sigma}}{v} \right)^2 \partial_\mu \chi \, \partial^\mu \chi - V(\bar{\sigma}) \tag{6.49}$$

with

$$V(\bar{\sigma}) = \lambda \, v \bar{\sigma}^3 + \frac{\lambda}{4} \bar{\sigma}^4 \,. \tag{6.50}$$

In this form, the original symmetry is still visible. Because only the derivative of the field χ enters, the shift of the field by a constant does not change the Lagrangian. The symmetry is hidden in the spectrum because the two fields do not have the same mass. The symmetry is also hidden in the Lagrangian of equations (6.43) and (6.44), where we used a poor choice of field variables. But it is really still there as can be seen using this better choice of field variables. That is the origin of Goldstone's theorem: the symmetry remains and the transformation in the symmetry direction does not change the energy. The shift symmetry prevents an $m^2 \chi^2$ term.

It is pretty remarkable how the different choices of field names described in the previous paragraph yield resulting Lagrangians that appear so different. By now you have had experience looking at Lagrangians and reading out their consequences via the Feynman rules. As we start to do this for these two Lagrangians, some things seem the same. For instance, the masses of the σ and $\bar{\sigma}$ are the same and both ϕ_I and χ are massless. So at the free field level these are the same theories. Their interactions, however, are completely different. The field ϕ_I has polynomial interactions without derivatives, while χ always has derivatives in its interactions. For example the $\sigma \phi_I \phi_I$ coupling that enters into the Feynman rules is

$$- 2i\lambda \, v \,, \tag{6.51}$$

while that of $\bar{\sigma} \chi \chi$ is

$$- 2i \frac{p_i \cdot p_f}{v} \,, \tag{6.52}$$

where p_i, p_f are the incoming and outgoing momenta of the χ fields. However, the surprising thing is that in the end the predictions of the two versions of the theory end up being the same. The names do not matter as long as the underlying physics is the same. To convince ourselves that this is true, let us evaluate the amplitude for

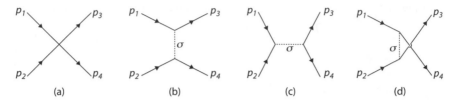

Figure 6.2. The Feynman diagrams for the scattering of massless fields. In diagrams (b), (c), and (d) the exchanged particle is a massive σ.

the scattering of the massless particles. Using the σ and ϕ_I variables we obtain the diagrams in figure 6.2, with the results

$$-i\mathcal{M} = -6i\lambda + (-2i\lambda v)^2 \left(\frac{i}{s - m_\sigma^2} + \frac{i}{t - m_\sigma^2} + \frac{i}{u - m_\sigma^2} \right), \qquad (6.53)$$

where $s = (p_1 + p_2)^2$, $t = (p_1 - p_3)^2$, and $u = (p_1 - p_4)^2$. These variables, for particles of mass m, satisfy $s + t + u = 4m^2$, which implies that this sum vanishes with massless particles such as we are using in the present example.

However, there is a series of cancellations in this amplitude. The first can be seen by using $2\lambda v^2 = m_\sigma^2$ and rewriting this amplitude as

$$-i\mathcal{M} = -6i\lambda - 2i\lambda m_\sigma^2 \left(\frac{1}{s - m_\sigma^2} + \frac{1}{t - m_\sigma^2} + \frac{1}{u - m_\sigma^2} \right)$$

$$= -2i\lambda \left(\frac{s}{s - m_\sigma^2} + \frac{t}{t - m_\sigma^2} + \frac{u}{u - m_\sigma^2} \right). \qquad (6.54)$$

This can be further simplified by adding $-2i\lambda (s + t + u)/m_\sigma^2$ to the amplitude, which is 0 by the kinematic relation, and combining terms such as

$$\frac{s}{s - m_\sigma^2} + \frac{s}{m_\sigma^2} = \frac{s^2}{m_\sigma^2 (s - m_\sigma^2)}. \qquad (6.55)$$

This leaves the amplitude as

$$-i\mathcal{M} = -2i\frac{\lambda}{m_\sigma^2} \left(\frac{s^2}{s - m_\sigma^2} + \frac{t^2}{t - m_\sigma^2} + \frac{u^2}{u - m_\sigma^2} \right). \qquad (6.56)$$

The point of this calculation is to show that the scattering amplitude is the same for both forms of the Lagrangian. We will now leave the final step in this proof up to you: use the Lagrangian of equation (6.49) to calculate the scattering amplitude and show that it is the same as equation (6.56).

While the details are left to you, we can see the important features in advance. There no longer is any quartic interaction, so that figure 6.2(a) does not exist. The remaining diagrams with the massive pole are the ones that count. Here each vertex is proportional to two factors of the momentum, and $2p_1 \cdot p_2 = (p_1 + p_2)^2$ for massless particles, so that we easily see the s^2, t^2, and u^2 factors. In the end, this way of calculating the amplitude is much more direct. The equality of the two calculations feels a bit remarkable, as even the diagrams involved are not the same. But it illustrates an important but subtle feature of Quantum Field Theory: field redefinitions preserve the physical predictions of the theory.

6.4 Example: Higgs mechanism and the Meissner effect

With a modest extension, the results of section 6.3 can illuminate some important phenomena. We do this by adding electromagnetism to the system. Recall from section 5.7 that the complex scalar field can describe a particle with an electromagnetic charge. This is accomplished by using the *covariant derivative*

$$\partial_\mu \phi \to D_\mu \phi = \left[\partial_\mu + ieA_\mu\right]\phi. \tag{6.57}$$

This slight addition makes an enormous change in the physical system.

If we identify the ground state as in section 6.3 and use the fields of equation (6.49), we have the Lagrangian

$$\mathcal{L} = -\frac{1}{4}F_{\mu\nu}F^{\mu\nu} + \frac{1}{2}\left(\partial_\mu\bar{\sigma}\partial^\mu\bar{\sigma} - m_\sigma^2\bar{\sigma}^2\right)$$
$$+ \frac{1}{2}\left(\frac{v+\bar{\sigma}}{v}\right)^2(evA_\mu + \partial_\mu\chi)^2 - V(\bar{\sigma}). \tag{6.58}$$

The A_μ term in the second line is the main new addition, following simply from the introduction of the covariant derivative. But now some magic happens. The field χ can actually be removed completely from the theory, by defining a new field

$$A'_\mu = A_\mu + \frac{1}{ev}\partial_\mu\chi. \tag{6.59}$$

We note that, thanks to the gauge invariance of the field strength tensor, we have $F'_{\mu\nu} = F_{\mu\nu}$, where $F'_{\mu\nu}$ is the field strength tensor formed with A'_μ. The Lagrangian becomes

$$\mathcal{L} = -\frac{1}{4}F'_{\mu\nu}F'^{\mu\nu} + \frac{1}{2}\left[\partial_\mu\bar{\sigma}\partial^\mu\bar{\sigma} - m_\sigma^2\bar{\sigma}^2\right] + \frac{1}{2}\left(\frac{v+\bar{\sigma}}{v}\right)^2 e^2v^2 A'_\mu A'^\mu - V(\bar{\sigma}). \tag{6.60}$$

Reading this version of the Lagrangian, we see that the photon has acquired a mass

$$m_\gamma^2 = e^2v^2. \tag{6.61}$$

This is quite a change from the result in section 6.3. The Goldstone boson has disappeared. The number of degrees of freedom has not changed because there is now an extra degree of freedom in the "photon" field—a massive spin-one particle has three spin states while the massless photon has only two polarizations. This is what is referred to as the Higgs mechanism (although Anderson, Brout, Englert, and others share credit for the insight).[1] In the Standard Model of particle physics, the Higgs mechanism is applied to the weak interaction gauge bosons rather than to the photon, but the implementation is just a generalization of what we have seen here.

This system can also exhibit the *Meissner effect*. Inside a superconductor the superconducting ground state "screens" the propagation of electromagnetic fields. This is equivalent to the Higgs solution of the previous paragraph.

However, in a strong external magnetic field, the normal (nonsuperconducting) ground state can be restored with no screening. We can simulate this by adding an external function $F_{\text{ext}}^{\mu\nu}$ to the Lagrangian in the form[2]

$$\mathcal{L}_{ext} = \mathcal{L} + \frac{1}{2} F_{\mu\nu} F_{\text{ext}}^{\mu\nu}. \tag{6.62}$$

In the presence of this term, the equations of motion, including a possible ground state value $\langle\phi\rangle$ for the scalar field, become

$$\partial_\mu \left(F^{\mu\nu} - F_{\text{ext}}^{\mu\nu} \right) + e^2 \langle\phi\rangle^2 A^\nu = 0. \tag{6.63}$$

If $\langle\phi\rangle = 0$, then the electromagnetic field is equal to this function $F^{\mu\nu} = F_{\text{ext}}^{\mu\nu}$, neglecting additional free field solutions, with $\partial_\mu F^{\mu\nu} = 0$, which always increase the energy. With $\langle\phi\rangle \neq 0$ there are more complicated solutions, but if the external field $F_{\text{ext}}^{\mu\nu}$ is constant, by looking for solutions with a uniform electromagnetic field, we force $A^\nu \to 0$, and hence $F^{\mu\nu}|_{\text{vac}} = 0$. The two possible solutions are thus the screened solution $\langle\phi\rangle \neq 0$, $F^{\mu\nu}|_{\text{vac}} = 0$ and unscreened one, $\langle\phi\rangle = 0$, $F^{\mu\nu}|_{\text{vac}} = F_{\text{ext}}^{\mu\nu}$.

Which of these solutions will be chosen? The ground state energy of these two solutions changes in the presence of the external field. Even the Hamiltonian changes, because the momentum conjugate to the field A_μ reads

$$\pi^\mu = \frac{\partial \mathcal{L}}{\partial(\partial_0 A_\mu)} = -F^{0\mu} + F_{\text{ext}}^{0\mu}, \tag{6.64}$$

such that

$$\mathcal{H} = \mathcal{H}_0 + F_{\text{ext}}^{0\mu} \dot{A}_\mu - \frac{1}{2} F_{\mu\nu} F_{\text{ext}}^{\mu\nu}. \tag{6.65}$$

If we specialize to an external magnetic field, this becomes

$$\mathcal{H} = \frac{1}{2} \mathbf{B}^2 - \mathbf{B} \cdot \mathbf{B}_{ext} + V(\phi). \tag{6.66}$$

---

[1]P. W. Anderson, "Plasmons, Gauge Invariance, and Mass," *Physical Review*, 1973, 130:439–442; P. W. Higgs, "Broken Symmetries, Massless Particles and Gauge Fields," *Physical Letters*, 1964, 12:132–133; F. Englert and R. Brout, "Broken Symmetry and the Mass of Gauge Vector Mesons," *Physical Review Letters*, 1964, 13:321–323; G. S. Guralnik, C. R. Hagen, and T. W. B. Kibble, "Global Conservation Laws and Massless Particles," *Physical Review Letters*, 13:585–587.

[2]We now drop the prime symbol in the notation of equation (6.60).

We can now compare the energy of the two solutions. The ground state solution with $\langle \phi \rangle = v$ and $\mathbf{B} = 0$ yields the usual energy density

$$\mathcal{H} = -\frac{\lambda}{4} v^4, \tag{6.67}$$

while the solution $\langle \phi \rangle = 0$ and $\mathbf{B} = \mathbf{B}_{ext}$ has energy

$$\mathcal{H} = -\frac{1}{2} \mathbf{B}_{ext}^2. \tag{6.68}$$

For small values of the external field, the usual screening solution is preferred. However, for external fields beyond a critical value of

$$\mathbf{B}_c^2 = \frac{\lambda}{2} v^4 \tag{6.69}$$

we revert to the unscreened solution with a massless photon and the external field penetrates the system. This mimics the Meissner effect.

Chapter summary: Finally, we have reached our goal of being able to make predictions of physical observables, such as decay rates and cross sections, and also on how to compute potentials. We have seen how symmetries of the action can be spontaneously broken, as well as some of the physical effects of this phenomenon. This chapter is just a small sampling of the many applications of Quantum Field Theory. Hopefully at this stage, the descriptions of field theoretic processes that you see in papers and seminars make sense to you. While each calculation is somewhat different, you can see what went into defining the theory and making the prediction.

CHAPTER 7

Introduction to renormalization

We have seen how to make predictions for decay rates and cross sections. These have been calculated using various interaction Lagrangians and have been parameterized by the coupling constants of the theory. However, we are still missing a key concept. These theoretical calculations are valid for any value of the coupling constants, as long as they are small enough, but its application in Nature is only to be applied for a specific value of this coupling, determined by measurements. That is, we need to *normalize* the prediction by measuring the coupling constant. Then when we do a better calculation by including higher orders in perturbation theory, we need to do this again to *renormalize* the couplings. This practice turns out to be more significant than it initially sounds. We explore that process here.

It should be stated up front that the study of renormalization can become a highly technical and detailed subject. Ours is a very introductory treatment, which emphasizes some of the conceptual issues and is very light on the technical aspects. It is meant as a starting point for further study.

7.1 Measurement

In this section we address an important question of principle. This involves our knowledge of the parameters that we use in making predictions. Let us start by looking at Quantum Electrodynamics.

The cross section for the Coulomb scattering of an electron on a proton is calculated to be

$$\frac{d\sigma}{d\Omega} = \frac{4\,m_e^2\,\alpha^2}{q^4}\,, \tag{7.1}$$

where q^μ is the four-momentum transfer. The quantum field–theoretical calculation does not know the value of α a priori—the electric charge is an arbitrary parameter in the Lagrangian. Studying how electrons bend in the Coulomb field of the proton is one way to measure this parameter. From this reaction, we can extract

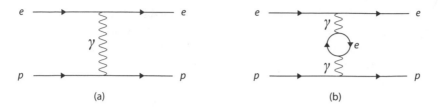

Figure 7.1. Coulomb scattering. Diagram (a) is the lowest order Feynman diagram, while (b) is the most important correction at the next order of perturbation theory.

the value of $\alpha = 1/137$ (the actual measured value[1] is $1/137.035999139(31)$, but in this discussion we will write $1/137$ for simplicity). This is the original normalization of the parameter α.

The leading correction to this is dominantly given by the diagram of figure 7.1(b), conventionally called $\Pi(q^2)$, with the result

$$\frac{d\sigma}{d\Omega} = \frac{4\, m_e^2\, \alpha_0^2 \left[1 + \Pi(q^2)\right]^2}{q^4}\,, \tag{7.2}$$

where we have temporarily put a subscript $_0$ on α in anticipation of the upcoming discussion. The issue is: what parameter α_0 should we use in this formula? Can we just insert $\alpha_0 = 1/137$? The answer is that we cannot use this value without some modification. The reason has to do with the way we measure α. While the quantum correction may modify the Coulomb potential at short distances, it should not change its strength at large distances, that is, where we measure α. Large distances correspond to the limit $q^2 \to 0$. If the quantum correction has $\Pi(0) \neq 0$, then what we have measured is

$$\alpha_{\text{phys}} = \alpha = \alpha_0\left[1 + \Pi(0)\right] = 1/137\,. \tag{7.3}$$

That is, we have had to *renormalize* α to account for our measurement of this parameter. Accounting for this, and recalling that $\Pi(q^2) = \mathcal{O}(\alpha)$, we can use equation (7.3) in the form $\alpha_0 \simeq \alpha\left[1 - \Pi(0) + \mathcal{O}(\alpha^2)\right]$, so that we can write

$$\frac{d\sigma}{d\Omega} = \frac{4\, m_e^2\, \alpha^2 \left[1 + \Pi(q^2) - \Pi(0) + \mathcal{O}(\alpha^2)\right]^2}{q^4}\,, \tag{7.4}$$

and here we can really use $\alpha = 1/137$. This is the process of renormalization. The renormalization program consists of first identifying how you intend to measure the parameter in question, then performing the measurement, and finally expressing the result in terms of the measured value.

---

[1] *Review of Particle Properties*, which is maintained and updated regularly by P. A. Zyla et al. for the Particle Data Group. See also Particle Data Group et al., "Review of Particle Physics," *Progress of Theoretical and Experimental Physics* 2020, no. 8 (August 2020): 083C01.

Let us explore this a bit further in the $\lambda\phi^4$ theory. This theory can be used for a short-range scattering process and the strength of the scattering depends on the specific process that we are planning to study. It is then clear that we have to determine the effective value of the parameter λ that suits the specific process. One way to do that it is to study the amplitude for $\phi\phi$ scattering. The cross section at lowest order was calculated in chapter 6 to be

$$\frac{d\sigma}{d\Omega} = \frac{|\mathcal{M}(s,\,t,\,u)|^2}{32\pi^2\,s}, \tag{7.5}$$

where $s = (p_1 + p_2)^2$, $t = (p_1 - p_3)^2$, and $u = (p_1 - p_4)^2$ are the kinematic variables, satisfying $s + t + u = 4\,m^2$. In particular, s is the square of the center of mass energy. At the lowest order in the coupling we have found

$$\mathcal{M}(s,\,t,\,u) = \lambda, \tag{7.6}$$

which implies that, to lowest order, the cross section reads

$$\frac{d\sigma}{d\Omega} = \frac{\lambda^2}{32\pi^2 s}. \tag{7.7}$$

There is a clear prediction here—the cross section scales inversely with the value of s, and it does not have any angular dependence in the center of mass. The cross section's magnitude, however, is not yet a prediction. We need some way to determine the experimentally appropriate value of the coupling λ. To do this we need to measure the cross section. We could do this near threshold, defined by $s = 4\,m^2$, $t = 0$, and $u = 0$, using

$$\lambda_{\mathrm{th}} = \mathcal{M}(4\,m^2,\,0,\,0), \tag{7.8}$$

naming the value that we measure near this point λ_{th}. Equivalently we could choose some other energy to measure the coupling. Let us imagine a measurement at $s = \mu^2$, $t = u = (4m^2 - \mu^2)/2$, and call the resulting value λ_μ. If the first-order formula was completely exact, these would yield the same value. Even if there is a small difference because we are using the first-order approximation of a perturbative series, either value is appropriate to use at this order in perturbation theory.

Now let us calculate the amplitude to the next order. Using the Feynman rules, we would generate the diagrams of figure 7.2. Each of the loop integrals depends on the momentum flowing through it. For example, figure 7.2(b) would be described by integral

$$I(s) = -i \int \frac{d^4 k}{(2\pi)^4} \frac{1}{k^2 - m^2 + i\epsilon} \frac{1}{(k + p_1 + p_2)^2 - m^2 + i\epsilon} \tag{7.9}$$

with $s = (p_1 + p_2)^2$. For our purposes at the moment, we need not calculate it, although we note simply from dimensional analysis that the integral is dimensionless. In terms of this integral, the matrix element is now

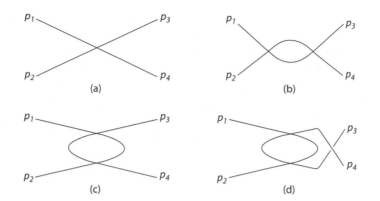

Figure 7.2. The diagrams that enter the matrix element for $\phi\phi$ scattering at one-loop order.

$$\mathcal{M}(s,t,u) = \lambda \left[1 - \frac{\lambda}{2} \left(I(s) + I(t) + I(u) \right) \right]. \tag{7.10}$$

How shall we now employ this new result? It would be incorrect to simply use the value of λ measured using the lowest order formula. Instead we clearly need to readjust the value of the coupling that we are using. There are various possibilities to do so. One simple choice would be to continue to use the same relation as at lowest order:

$$\mathcal{M}(4\,m^2,\, 0,\, 0) = \lambda_{\text{th}}^{\text{ren}}, \tag{7.11}$$

where the superscript ^{ren} indicates that this is a new renormalized value. If we do this, the matrix element to this order in perturbation theory becomes

$$\mathcal{M}(s,\, t,\, u) = \lambda_{\text{th}}^{\text{ren}} \left[1 - \frac{\lambda_{\text{th}}^{\text{ren}}}{2} \left(I(s) - I(4\,m^2) + I(t) + I(u) - 2\,I(0) \right) \right]. \tag{7.12}$$

Once we are provided with the functional form of $I(s)$, this becomes a more accurate description of the scattering reaction.

At this order we would not get the same value for the coupling if we repeated this procedure at a different kinematic point. If we measured at $s = \mu^2$, $t = u = (4m^2 - \mu^2)/2$, using

$$\lambda_{\mu}^{\text{ren}} = \text{Re} \left\{ \mathcal{M}(\mu^2, (4m^2 - \mu^2)/2, (4m^2 - \mu^2)/2) \right\}, \tag{7.13}$$

the coupling would differ slightly.[2] However, the matrix element

$$\mathcal{M} = \lambda_{\mu}^{\text{ren}} \left[1 - \frac{\lambda_{\mu}^{\text{ren}}}{2} \left(I(s) + I(t) + I(u) - \text{Re} \left\{ I(\mu^2) + 2\,I((4m^2 - \mu^2)/2) \right\} \right) \right] \tag{7.14}$$

---

[2]We have used the real part of the matrix element because, as we will see, the loop integrals can have imaginary components when $p^2 > 4\,m^2$.

would be equally valid to this order in the calculation as is our previous expression, equation (7.10).

Up to this point, we have used on-shell kinematics ($s + t + u = 4\,m^2$) in defining the coupling, but the renormalization point chosen need not even be a physical value of the momenta. For example, symmetry and aesthetics might lead us to define the coupling using the symmetric point $s = t = u = 0$, which is not compatible with all the external particles being on-shell. However, that does not matter because we have the loop integral defined at all momenta. We measure in the physical region and use the formula to refer back to the unphysical renormalizaton point. In this case we would have

$$\mathcal{M} = \lambda_{\mathrm{sym}}^{\mathrm{ren}} \left[1 - \frac{\lambda_{\mathrm{sym}}^{\mathrm{ren}}}{2} (\Delta I(s) + \Delta I(t) + \Delta I(u)) \right] \tag{7.15}$$

with $\Delta I(p^2) \equiv I(p^2) - I(0)$. This is probably the simplest variation of those that we have mentioned. In the appendix we calculate

$$I(p^2) - I(0) = -\frac{1}{16\pi^2} \int_0^1 dx \, \log \left(\frac{m^2 - p^2 x(1-x) - i\epsilon}{m^2} \right). \tag{7.16}$$

Use of this formula would allow us to measure the coupling $\lambda_{\mathrm{sym}}^{\mathrm{ren}}$. The output value of the coupling λ would be slightly different in each of these cases, although the different definitions could be calculably related to this order in perturbation theory. The important thing is that we have a well-defined formula to use in the measurement.

Clearly this could be repeated at many different *renormalization points*, such as at different values of μ. In addition, there could be different *renormalization schemes* from the ones that we used, as we do not have to use the exact identifications posed in this section when doing the measurement. Finally, the procedure must be repeated if we improve the calculation by including the next order in perturbation theory. But the basic point is that when we measure the coupling constant in formulas with higher order interactions, we need to give a precise specification, and then reexpress the matrix element in terms of that specified measured value.

7.2 Importance of the uncertainty principle

This is a good place to raise an important question. How can we make any predictions in quantum mechanics or Quantum Field Theory if we do not know the fundamental theory at all energies? Physics is an experimental science, and we have explored it over a range of energies. As we have proceeded to higher energies we have found new particles and new interactions. This by itself is expected, but it does raise an issue about theoretical calculations. In quantum mechanics we are instructed to sum over *all* intermediate states I

$$\mathcal{M}_{fi} = \sum_I \frac{\langle f|V|I\rangle \langle I|V|i\rangle}{E - E_I}. \tag{7.17}$$

How can we do this if there might be some intermediate states that we have not discovered yet? Moreover, in addition to possible new particles, we have loops of all energies and we do not know for sure that the high momentum behavior of the loops is correct, as we have not yet probed those energies. We will see in section 7.3 and in the appendix, where we explicitly calculate loop diagrams, that high mass particles and high energies in loops *can* contribute to amplitudes. How can we make predictions at all if we do not know the physics?

The answer is essentially given by the uncertainty principle. Effects from very high energy appear local—that is, confined to short distances—to us who live at low energy. When we do perturbation theory in quantum mechanics, we say that we can violate energy only for a short time of order $\Delta t \sim 1/\Delta E$. In Quantum Field Theory, propagators of heavy particles only propagate for a short distance, and the high momentum components of loops are short distance dominated because of the uncertainty principle. So for both tree diagrams and loops, we see the effects of very high energy as local interactions.

In Quantum Field Theory local interactions are described by the interaction terms in the Lagrangian. Each of these comes with a coefficient that we have been calling a coupling constant. The unknown effects from high energy, along with known effects, will show up as shifts in these parameters. As long as we are general enough in constructing our interaction Lagrangians, there will be a local term relevant for every possible high energy effect. But, as we have discussed in section 7.1, we do not know these couplings in advance. We have to go and measure them. In doing so, we lose track of how large was the "bare" parameter and how large was the shift due to the physics from high energies. We only see the overall renormalized value that was measured in some specific way. The unknown high energy physics is contained as some part of the measured value of the parameter.

This is very fortunate as it allows predictions when we do not know the physics at all high energies. We are protected from needing to know these effects by the uncertainty principle and the renormalization procedure.

7.3 Divergences

One of the widely known features of Quantum Field Theory is that it is plagued by divergences. This has traditionally been cited as the need for renormalization. We have kept this aspect hidden until now to emphasize that renormalization is needed even if the theory is finite: divergences are not the cause for this procedure. Moreover, as we have just emphasized in section 7.2, there are many unknowns at high energy because we have not yet explored all energies. Divergences are just one of these uncertainties. We have no way of knowing whether the divergences really occur or whether there is some compensating physics that makes a finite theory. But in the end, these unknowns do not matter. Because all unknown high energy physics is local when viewed at low energy, and because we measure the experimental values of the low energy constants as required by renormalization, all ultraviolet divergences are irrelevant for physics at ordinary energies.

Nevertheless, it is important to know some of the facts about divergences. Let us go back to the loop integral in the scattering amplitude in ϕ^4 theory. It is

$$I(p^2) = -i \int \frac{d^4k}{(2\pi)^4} \frac{1}{k^2 - m^2 + i\epsilon} \frac{1}{(k-p)^2 - m^2 + i\epsilon} . \qquad (7.18)$$

This integral is dimensionless. A look at the powers of momentum suggests (correctly) that it goes as

$$I \sim \int \frac{dk}{k} \qquad (7.19)$$

at high momentum, which implies that it is logarithmically divergent. Rather than dealing with divergent functions, theorists will modify the full behavior in some way to make the integral finite, a process that is referred to as *regularization*. At the end of the calculation the regularization can be removed. Common ways to regularize this integral are a high energy cutoff on the momentum or dimensional regularization, which involves continuing the integral from four to d dimensions.

In the appendix, we show how to evaluate this integral with a cutoff Λ at high energy. The result is

$$I(p^2) = \frac{1}{16\pi^2} \left[\log(\Lambda^2/m^2) - \int_0^1 dx \log \left(\frac{m^2 - p^2x(1-x)}{m^2} \right) - 1 + O(\Lambda^{-2}) \right].$$
$$(7.20)$$

In dimensional regularization the equivalent quantity is

$$I(p^2) = \frac{1}{16\pi^2} \left[\frac{2}{4-d} + c + \log(\mu^2/m^2) - \int_0^1 dx \log \left(\frac{m^2 - p^2x(1-x)}{m^2} \right) \right.$$
$$\left. + O(d-4) \right], \qquad (7.21)$$

where c is a numerical constant that we will not care about and μ is an arbitrary energy scale that must be introduced when using dimensional regularization to keep everything dimensionally consistent. We see that

$$I(0) = \frac{1}{16\pi^2} \left[\log(\Lambda^2/m^2) - 1 \right] \quad \text{or} \quad \frac{1}{16\pi^2} \left[\frac{2}{4-d} + c + \log(\mu^2/m^2) \right] \quad (7.22)$$

and also recover the expression for $\Delta I(p^2) = I(p^2) - I(0)$ from equation (7.16), which is finite and independent of the regularization scheme. From this we can see that all of the different forms of renormalized couplings, which were defined by this procedure, are finite. The procedure we defined for measuring coupling constants also has the effect of removing the divergences.

At this stage we can get an insight into why renormalization should be universal, that is, valid in all processes. All high energy effects are equivalent to a shift in the coefficient of a local term in a Lagrangian. The high momentum behavior of loops is one such effect. In the Appendix we will show that the divergence that we have found is equivalent to a Dirac delta function in coordinate space. It therefore gets hidden during the renormalization process as it is one part of the measured value. As long as every possible divergence has a parameter to be absorbed into—and on general principles it should—then all divergences can sort themselves into the various possible renormalized couplings. This heuristic argument underrepresents the real complexity of proving renormalizability of all divergences, especially that of constructing proofs that are valid to all orders in perturbation theory. However, it does capture the essential physics of why divergences do not disrupt our predictive power.

So far we have focused on the renormalization of the coupling constants, but there is also a need for renormalization of the mass and of the wavefunction of the fields. The first is reasonably intuitive. If there are quantum corrections that look like a mass term, we need to adjust the total to yield the measured value of the mass. Wavefunction renormalization is a little less intuitive. However, it arises because interactions can change the normalization of the states, which follows from the commutation rules. This needs to be readjusted back to the canonical value. We address these issues in section 7.4.

What would happen if there was extra new physics in this process? For example, if there were an extra very heavy particle that was coupled to the light scalars. This could be a field χ with a big mass M coupled to ϕ with the Lagrangian

$$\mathcal{L} = -\frac{\bar{\lambda}}{4}\phi^2\chi^2. \tag{7.23}$$

As a practice with reading Lagrangians, you can show that the Feynman diagram vertex for this case is $-i\bar{\lambda}$. The contribution to the scattering process is the same as in equation (7.10) with $\lambda^2 I(p^2)$ replaced by $\bar{\lambda}^2 \bar{I}(p^2)$, where the latter is the same integral in equation (7.18) but with m^2 replaced by M^2. So we see that the contribution of this heavy field is not small; it is formally divergent. Moreover, even if you throw away the divergent term, the remainder depends on the mass M^2 of the heavy particle (as can be seen in equations (7.20) and (7.21)), so even the finite piece is present. However, all of this goes into the renormalization of λ. Once we measure it we have differences that, for $|p^2| \ll M^2$, have the form

$$\bar{I}(p^2) - \bar{I}(0) = \frac{1}{96\pi^2}\frac{p^2}{M^2} + O(p^4/M^4), \tag{7.24}$$

which is highly suppressed for large M. So we see that the potentially large direct contribution is absorbed into the local $\lambda\phi^4$ coupling, and the residual is suppressed by powers of the heavy scale M. Indeed this contribution can itself be described by a local operator, in particular

$$\bar{\mathcal{L}} = \frac{\bar{\lambda}^2}{768\pi^2}\phi^2\frac{\Box}{M^2}\phi^2. \tag{7.25}$$

This operator has been defined to have the same matrix element for the scattering amplitude as we found by evaluating the loop in the heavy mass limit.

Let us summarize the philosophy here. We know nothing about the true high energy behavior of our theories, because we have not been able to explore the particles and interactions at such energies. Quantum Field Theory indicates that there are divergences, but these divergences may well disappear in the full, final theory. However, in the end it does not matter what happens in the ultraviolet. This is because the high energy effects appear local at low energy. Their effect is to shift the parameters of local terms in the Lagrangian. Because we need to measure these parameters, the heavy effects are unobservable. Residual corrections can appear as higher dimensional operators with coefficients that are suppressed by powers of the heavy mass.

7.4 Techniques

Thus far, our discussion of the need for renormalization does not use the optimal procedures for actual calculations. Now that we understand a bit more where we are headed, we can look at a better strategy.

We now recognize that the parameters in our original Lagrangian are not the renormalized parameters that we measure. However, in the end we do want to use the physical parameters to express our final answer. We can do this by rewriting the Lagrangian in terms of the physical parameters and *counterterms*. To see how this works, let us write our Lagrangian initially in terms of *bare parameters*, that is, of parameters m_0 and λ_0 that appear in the calculations, but which are not the observed quantities, and with an arbitrary normalization Z of the field:

$$\mathcal{L} = \frac{1}{2} Z(\partial_\mu \phi \partial^\mu \phi - m_0^2 \phi^2) - \frac{\lambda_0 Z^2}{4!} \phi^4 . \qquad (7.26)$$

This can be converted to the correct normalization and measured values m, and λ by rewriting it as

$$\mathcal{L} = \frac{1}{2} (\partial_\mu \phi \partial^\mu \phi - m^2 \phi^2) - \frac{\lambda}{4!} \phi^4 + \Delta \mathcal{L}_{\text{ct}} , \qquad (7.27)$$

with the counterterm Lagrangian being

$$\Delta \mathcal{L}_{\text{ct}} = \frac{1}{2} \Delta Z(\partial_\mu \phi \partial^\mu \phi - m^2 \phi^2) - \frac{1}{2} \Delta m^2 \phi^2 - \frac{1}{4!} \Delta \lambda \, \phi^4 \qquad (7.28)$$

and $\Delta Z = Z - 1$, $\Delta m^2 = Z(m_0^2 - m^2)$, and $\Delta \lambda = Z^2 \lambda_0 - \lambda$. These counterterms can now be treated as new interactions and adjusted to whatever value is needed to satisfy the renormalization conditions.

For example, if we want to choose an on-shell renormalizaton condition, we would adjust the counterterms so that the pole in the propagator is located at the value of the physical mass. The propagator is modified by self-energy shown in figure 7.3. The solid circle is the set of Feynman diagrams that can modify the

Figure 7.3. The corrections to a propagator. The box signifies the sum of two contributions. The self-energy $\Sigma(q)$ is represented by the solid circle, while the $\times$ represents the counterterms defined in equation (7.28).

Figure 7.4. The summation of self-energy corrections to the propagator. The box is defined in the diagram of figure 7.3.

one-particle line and is traditionally denoted by $\Sigma(q)$. The $\times$ represents the counterterms, that are giving a Feynman diagram "vertex" $-i(\Delta Z[q^2 - m^2] - \Delta m^2)$. Repeated use of this self-energy occurs in the diagrams shown in figure 7.4. These can be summed using the geometric series

$$\frac{1}{A - B} = \frac{1}{A} + \frac{1}{A}B\frac{1}{A} + \frac{1}{A}B\frac{1}{A}B\frac{1}{A} + \dots \tag{7.29}$$

to yield the propagator

$$iD_F(q) = \frac{i}{q^2 - m^2 - \left(\Sigma(q) + \Delta Z[q^2 - m^2] - \Delta m^2\right)}, \tag{7.30}$$

using the geometric sum represented in figure 7.4. For on-shell renormalization, we note that in general $\Sigma(q)$ can be expanded about the physical mass with the form

$$\Sigma(q) = a + b\,(q^2 - m^2) + c\,(q^2 - m^2)^2 + \dots, \tag{7.31}$$

where a, b, and c are constants. If we choose the counterterms correctly, we can cancel the first two terms of this expansion, and the propagator then has a pole in the correct location and with the correct normalization.

This explains why we are able to drop the self-energy diagrams on external lines when applying the Feynman rules, as noted in section 5.4. Both the self-energy and the counterterm appear on the external line in the same combination as in diagram (5.33). They will cancel by the on-shell renormalization condition.

The same story holds for the coupling constant. When we include the counterterm in the perturbative expansion, we now have the matrix element

$$\mathcal{M}(s, t, u) = \left[\lambda + \Delta\lambda - \frac{\lambda^2}{2}(I(s) + I(t) + I(u))\right], \tag{7.32}$$

and the counterterm $\Delta\lambda$ can be chosen to impose whatever renormalization condition we want. Now that these counterterms are defined, they must be consistently used in whatever other processes we choose to calculate.

7.5 The renormalization group

Imagine that you have measured the renormalized coupling constant for the $\phi\phi$ scattering at threshold and are now studying the reaction at much higher energies s, $|t|$, and $|u| \gg m^2$. In this regime, there are logarithmic corrections that, using equation (7.16), can be written in the form

$$
\mathcal{M} \simeq \lambda_{\text{th}}^{\text{ren}} \left[1 + \frac{\lambda_{\text{th}}^{\text{ren}}}{32\pi^2} \left(\log\left(-\frac{s}{m^2}\right) + \log\left(-\frac{t}{m^2}\right) + \log\left(-\frac{u}{m^2}\right) + \text{const.} \right) \right].
$$
(7.33)

We see that the magnitude of this matrix element grows larger at high energy. The logarithms can become arbitrarily large and could appear to overwhelm the perturbative expansion. The culprit appears to be the fact that we initially renormalized the coupling at threshold $s = 4\,m^2$, and we are now applying the result at kinematics far from this energy. However, repackaging this in terms of a *running coupling constant* can help remove large logarithms and lead to an expansion that is more reasonable. Let us instead measure the coupling at a higher kinematic point comparable to the energies that we are now using, $s = \mu_1^2 \gg 4\,m^2$, and $t = u = -\mu_1^2/2$. We did this in equation (7.11), except that here we are neglecting the particle mass compared to the energy scale. The matrix element now becomes

$$
\mathcal{M} = \lambda(\mu_1) \left[1 + \frac{\lambda(\mu_1)}{32\pi^2} \left(\log\left(\frac{s}{\mu_1^2}\right) + \log\left(\frac{t}{\mu_1^2}\right) + \log\left(\frac{u}{\mu_1^2}\right) + \text{const.} \right) \right].
$$
(7.34)

The large logarithms are no longer present,[3] and $\lambda(\mu_1)$ is a more appropriate coupling constant to use at these energies because it more accurately describes the physics at these scales.

If we change scales, the effective coupling will change. If we had used a different scale μ_2, the couplings would be related by

$$
\lambda(\mu_2) = \lambda(\mu_1) + \frac{3}{32\pi^2}\lambda^2(\mu_1) \log\frac{\mu_2^2}{\mu_1^2}.
$$
(7.35)

We can readily convert this into a differential equation

$$
\mu\frac{d}{d\mu}\lambda(\mu) = \frac{3}{16\pi^2}\lambda^2(\mu) = \beta(\lambda),
$$
(7.36)

---

[3]We could have used an off-shell symmetric renormalization point $s = t = u = \mu_1^2$, and it would have worked equally well. The logarithmic terms would be the same and the small additional constant would change slightly.

which is called the renormalization group equation for the coupling. It also identifies the so-called beta function $\beta(\lambda)$. Equation (7.36) has, as a general solution,

$$\lambda(\mu_1) = \frac{16\pi^2}{3\log\frac{\Lambda^2}{\mu_1^2}}, \tag{7.37}$$

where Λ is an integration constant, because this is a first-order differential equation. The constant Λ can be determined numerically by measuring λ at some specified value $\mu_1 \gg m$. If, as is required by the validity of perturbation theory, that measurement yields $\lambda(\mu_1) \ll 1$, then Λ will found to be $\Lambda \gg \mu_1$.

This is our running coupling constant that is dependent on the renormalization scale μ. It captures the important quantum corrections appropriate for that scale and is the best representation of the coupling strength near that energy scale.

The idea of scale-dependent coupling constants has widespread applicability. In analyzing condensed matter systems, it is useful to start with the short distance microscopic theory and transform to longer distance scales. In particle physics it is often the reverse—scaling from low to high energies as the energy scale of experiments grows.

7.6 Power counting and renormalization

In section 4.4 we showed how the dimensionality of the fields in a given term in the Lagrangian determines the dimension of the coupling constant that parameterizes the strength of that interaction. For example, in a $\lambda\phi^4$ interacton, λ would be dimensionless because the four scalar fields add up to mass dimension four. This already tells us something useful about renormalization.

An example will help explain this. For the purposes of this example we can treat the scalar fields as massless and use dimensional regularization. Neither of these two choices is really required, but they are useful because in this case the only dimensional factors in loop diagrams are the momenta. Once the loop momentum is integrated over, the overall dimension of the integral must be given by factors of the external momenta.

In the $\phi\phi$ scattering process discussed at the start of this chapter the original coupling λ was modified by loops of order λ^2. The divergent part in dimensional regularization is always proportional to $2/(4-d)$. The one-loop correction to the matrix element for $\phi\phi$ scattering then is equivalent to the shift

$$\lambda \to \left[\lambda + \lambda^2\left(\frac{2c_1}{4-d} + \dots\right)\right] \sim \lambda^{\text{ren}}, \tag{7.38}$$

where c_1 is some dimensionless constant. This then is clearly just a shift in λ and the divergence goes into the renormalized value of the coupling. At higher orders of perturbation theory there will be higher powers of λ, but the coefficients appearing in the relationship between bare and renormalized parameters will always be dimensionless.

Now replace the operator $\lambda\phi^4$ we used to describe the $\phi\phi \to \phi\phi$ scattering with one with more derivatives, for example,

$$\mathcal{L}' = \frac{1}{2M_*^4} (\partial_\mu\phi\partial^\mu\phi)^2 \,. \tag{7.39}$$

By counting the dimensions of derivatives and fields we see that the coupling constant in the Lagrangian has to have dimension of M^{-4}, so we have named the coupling constant here $1/(2M_*^4)$, where M_* has dimensions of a mass. The tree-level matrix element is now

$$\mathcal{M} = -\frac{1}{M_*^4}(s^2 + t^2 + u^2) \,. \tag{7.40}$$

As before, this matrix element is dimensionless, although now there are factors of momentum in the numerator. When calculating the one-loop corrections to this, there will be two factors of the coupling constant, which means that there is an overall factor of $1/M_*^8$. Because we have taken the scalars to be massless and we are using dimensional regularization, the numerator must involve the external momentum, so that the matrix element, at one-loop order, will get a correction

$$\delta\mathcal{M}_{1\ loop} \sim \frac{1}{M_*^8}(s^4 + \dots) \,. \tag{7.41}$$

Specifically, some work shows that the divergent terms are

$$\frac{1}{48\pi^2 M_*^8}\left[(s^4 + t^4 + u^4) - \frac{1}{5}(s^2 + t^2 + u^2)^2\right]\left(\frac{2}{4-d} + \dots\right) \,. \tag{7.42}$$

These terms have a different momentum dependence than the original interaction does. This implies that these divergences cannot be absorbed into the original coupling constant. Instead they are like new terms in the Lagrangian with higher derivatives, such as

$$\bar{\mathcal{L}} = \frac{1}{4\bar{M}^8}(\partial_\mu\phi\partial^\mu\phi)\Box^2(\partial_\nu\phi\partial^\nu\phi) + \frac{1}{4\hat{M}^8}(\partial_\mu\phi\partial^\nu\phi)\Box^2(\partial_\mu\phi\partial^\nu\phi) \,, \tag{7.43}$$

with new coupling constants that we have called $1/(4\bar{M}^8)$ and $1/(4\hat{M}^8)$ because they carry the dimensionality of M^{-8}. Now we are able to absorb the divergence above into this coupling. But unfortunately that is not the end of the story. When we go to two loops we get yet new divergences, with twelve powers of momenta. Likewise using the operators of equation (7.43) in loops we get yet more divergences with higher order momentum dependence. To renormalize these, we need to define yet more operators of yet higher dimension. This story then continues ad infinitum.

This illustrates the concept of "renormalizable" versus "nonrenormalizable" theories. In those with dimensionless couplings (or couplings with a positive mass dimension), the divergences can be absorbed into a finite (and typically small!) number of couplings. These are the renormalizable theories. This condition can be rephrased as one on the powers of fields that appear in the interaction Lagrangian. Renormalizable theories are those where the mass dimension of those interaction terms is ≤ 4. For theories whose couplings carry inverse powers of mass, or equivalently whose Lagrangians contain operators whose dimensions add up to > 4, the number of operators needed to absorb the divergences increases with each loop order. This are called nonrenormalizeable. The name is a bit unfortunate, as we can still renormalize the divergences in these theories. But it takes an infinite number of operators to do this to all orders in perturbation theory.

Renormalizable theories are more predictive. In a theory like $\lambda\phi^4$, only two parameters, the mass m and coupling λ, are needed to remove the divergences that occur at any loop order. Once we have measured their renormalized values in some scheme, we can predict all processes within the theory. Let us now discuss how also nonrenormalizable theories can be useful.

7.7 Effective field theory in brief

For many years it was thought that the only predictive quantum field theories were of the renormalizable variety. To fully identify a nonrenormalizable theory, we would need an infinite number of measurements to determine the coefficients of an infinite number of operators. However, we now understand that predictions can be made in at least some nonrenormalizable theories. Moreover, there are even situations when it is a good idea to convert a renormalizable theory into a simpler, but nonrenormalizable, version. This section expands on the logic of nonrenormalizable effective field theories.

The fundamental logic is that when we are working at low energies, we need to consider as explicit degrees of freedom only those particles that are active at those energies. If there are other heavier particles in the theory, they are not excited. As we argued in section 7.3, their effect could appear as a local operator. We gave an example of this in equation (7.25). So we may construct a low energy theory that has the light fields in it, along with more local operators that mimic the virtual effects of the high energy particle. However, the effective field theory is more than just the existence of these local operators: it is a full quantum field theory, and these new operators can also be used in loop processes.

We can explicitly construct an effective theory building on our previous work. Recall the example of spontaneous symmetry breaking discussed in section 6.3. There we started from the Lagrangian

$$\mathcal{L} = \partial_\mu \phi^* \partial^\mu \phi + \mu^2 \phi^* \phi - \lambda(\phi^*\phi)^2 \,. \tag{7.44}$$

After finding the ground state, with $v = \sqrt{\mu^2/\lambda}$, we defined new fields

$$\phi(x) = \frac{1}{\sqrt{2}}\,(v + \bar{\sigma}(x))\,e^{i\chi(x)/v}\,. \tag{7.45}$$

Figure 7.5. The interaction generated by the exchange of one quantum of $\bar{\sigma}$. The factors of $\times$ represent the interaction term $\frac{1}{v} \partial_\mu \chi \partial^\mu \chi$.

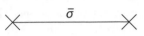

In terms of these fields, the Lagrangian becomes

$$\mathcal{L} = \frac{1}{2}(\partial_\mu \bar{\sigma} \partial^\mu \bar{\sigma} - m_\sigma^2 \bar{\sigma}^2) + \frac{1}{2}\left(\frac{v+\bar{\sigma}}{v}\right)^2 \partial_\mu \chi \, \partial^\mu \chi - V(\bar{\sigma}), \qquad (7.46)$$

where $m_\sigma^2 = 2\,\mu^2$ and where the potential

$$V(\bar{\sigma}) = \lambda\, v\, \bar{\sigma}^3 + \frac{\lambda}{4}\,\bar{\sigma}^4 \qquad (7.47)$$

contains only $\bar{\sigma}$ and not χ. All interactions of the Goldstone boson χ involve derivatives.

Because the $\bar{\sigma}$ field is heavy, we might be tempted to drop any reference to it when working at low energy (in this case, "low energy" means well below the mass of $\bar{\sigma}$). Here only the Goldstone boson is an active degree of freedom. However, this would leave us with only a free field theory without interactions. The interactions come from the coupling of χ to $\bar{\sigma}$, and there can be a virtual effect of the exchange of the heavy particle as in figure 7.5. Calculating the effect of this leads to

$$S_{\bar{\sigma}\ \text{exchange}} = -\frac{1}{2v^2} \int d^4x\, d^4y (\partial_\mu \chi \partial^\mu \chi)\,(x)\, D_F(x-y)\,(\partial_\nu \chi \partial^\nu \chi)\,(y)$$

$$\simeq \int d^4x\, \frac{1}{2\,v^2 m_\sigma^2}\,(\partial_\mu \chi \partial^\mu \chi)(x)(\partial_\nu \chi \partial^\nu \chi)(x). \qquad (7.48)$$

In writing this the remaining fields are meant to act on an external state, but there are no $\bar{\sigma}$ fields in the external states at this energy, so this field must be contracted to form the propagator. Because the $\bar{\sigma}$ mass is heavy, the position space propagator is well approximated by a Dirac delta function. The calculation results in an effective Lagrangian

$$\mathcal{L}_{\text{eff}} = \frac{1}{2}\,\partial_\mu \chi \partial^\mu \chi + \frac{1}{2\,v^2 m_\sigma^2}\,(\partial_\mu \chi \partial^\mu \chi)^2. \qquad (7.49)$$

We can readily check that this new interaction is equivalent to the nonrenormalizable interaction that we discussed in equation (7.39) with the identification

$$\frac{1}{2\,M_*^4} = \frac{1}{2\,v^2 m_\sigma^2}. \qquad (7.50)$$

We have thus converted a renormalizable field theory into an effective nonrenormalizable one. Note that given the identification of the coupling constants in equation (7.50), the scattering amplitude of equation (7.40) is the same as the tree-level scattering of the original theory given in equation (6.56) in the limit s, t, $u \ll m_\sigma^2$.

This confirms the equivalence of the effective theory to the original one at low energy. You will note that the calculation of the effective theory is considerably simpler.

However, one might be concerned that, because the effective field theory is technically nonrenormalizable, we might not make further predictions. When we discussed this interaction in section 7.6, the focus was on the disadvantages of non-renormalizable interactions, in that there is an increasing number of operators with more derivatives that are needed when performing higher order loops. Here we would like to turn this feature into an advantage. We will use the fact that derivatives in operators turn into factors of momentum when matrix elements are taken. Then when working at low momentum, operators with more derivatives have a smaller effect than those with fewer derivatives. In our present example, the momentum expansion is in powers of p^2/m_σ^2 and $p^4/v^2 m_\sigma^2$. When working to a given accuracy, one needs to keep only a subset of operators with the fewest derivatives. And because loops bring in more powers of the derivatives, we only need to perform calculations up to a small number of loop orders.

There is still the worry that if we were to use this effective interaction in loops, we would get the wrong answer. After all, the full theory also has loops involving the σ. And in fact the high energy behavior of any loop calculation using the effective Lagrangian *would* be inaccurate. However, that does not matter. These high energy effects appear as local at low energy and would be described by coefficients in a local Lagrangian. Because the effective field theory is being applied at low energy, what matters is the low energy part of loop diagrams. For this, only the χ particles are active, and the effective field theory describes the couplings and propagation of the light particles correctly. In practice this means that kinematic effects that are *not* equivalent to a local term in the action have come from low energy and have been correctly predicted.

When using effective field theory, we include only the light particles—those that are active at the energy under investigation. In general, we must write out the most general local Lagrangian for these particles that is consistent with the symmetries of the theory. This will include terms that are formally nonrenormalizable, but that can be ordered in an energy expansion by the dimensionality of the operators. The lowest dimension operators are the dominant ones at low energy. Most often the coefficients of these will be known either from a more complete theory or from measurement. Using these, we can start calculating.

The divergences are a boring part of the calculation. Because they come from high energy where the effective theory is not valid, they are not interesting. But they can be absorbed by defining renormalized values of the coupling using some prescription. These values themselves are more interesting, but they are not predicted by the effective field theory by itself. Either they can be calculated by matching to the full theory or, if that full theory is unknown or too complicated, they can be measured. At a given order this has reduced the effect of high energy to just a few numbers. There will be residual parts of the calculation that are dynamical predictions.

We can see this at work in $\chi\chi$ scattering. By the rules of effective field theory, we should include all operators consistent with the symmetry of the theory. In this case, the symmetry is a shift symmetry $\chi \to \chi + c$. These are then arranged in an expansion in the number of derivatives. We will not display all of these. As we saw in section 7.6, we will need some operators with eight derivatives in order to absorb the divergences. We choose these to be those of equation (7.43). Given these ingredients and some calculation, we find the matrix element to be

$$
\begin{aligned}
\mathcal{M} = &-\frac{1}{v^2 \, m_\sigma^2} \, (s^2 + t^2 + u^2) + \ldots \\
&+ \left(\frac{1}{4\,\hat{M}^8(\mu)} - \frac{1}{2\,\bar{M}^8(\mu)} \right) (s^4 + t^4 + u^4) - \frac{1}{4\,\hat{M}^8(\mu)} \, (s^2 + t^2 + u^2)^2 \\
&- \frac{1}{36\pi^2 \, v^4 \, m_\sigma^4} \left[\left(s^4 \log \frac{-s}{\mu^2} + t^4 \log \frac{-t}{\mu^2} + u^4 \log \frac{-u}{\mu^2} \right) \right. \\
&\left. - \frac{1}{5} \, (s^2 + t^2 + u^2) \left(s^2 \log \frac{-s}{\mu^2} + t^2 \log \frac{-t}{\mu^2} + u^2 \log \frac{-u}{\mu^2} \right) \right].
\end{aligned}
\tag{7.51}
$$

There are many features to note in this result. The first line contains the lowest order tree-level matrix element. The ellipses there represent the contributions from all the other operators, which we are not considering here. The second line contains the renormalized higher order operators that we have used to absorb the infinities of the calculation. By denoting them by $\bar{M}(\mu)$ and $\hat{M}(\mu)$, we recognize that their renormalized values will depend on the scale μ, which enters the formula on the next lines. These parameters are not predictions of the effective field theory, although if we worked hard enough we could determine them from the couplings of the original theory. Finally, the last two lines contain logarithmic corrections coming from the finite part of the one-loop contribution. Because they are nonanalytic in the momenta, they cannot come from a local term in a Lagrangian, so they come from the low energy parts of the loops. These are the true one-loop predictions of this theory.

Effective field theory can be used as the low energy limit of a known renormalizable field theory. In this case, it is used because it simplifies the calculations and focuses them on the important physics. The heavy particles are rarely important, but the low energy ones always are. Effective field theory can also be used in other settings where the full theory is unknown or difficult. For example, in the description of nucleons and pions, the full theory is Quantum Chromodynamics, which is very hard to deal with at low energies. If one were to insist on making a renormalizable field theory of nucleons and pions, that theory would be overly restrictive. It is actually the wrong theory and the restriction to normalizable interactions fails to capture the correct interactions. An effective field theory works well here, as it describes the most general pion-nucleon interactions consistent with the relevant symmetries. Likewise, in situations where we have no knowledge of the ultimate high energy theory, such as quantum gravity, the effective field theory treatment is the only reliable approach.

This discussion has necessarily been brief—merely an orientation to what has now become a widely used technique.

Chapter summary: Here we have seen some more advanced field-theoretical reasoning. This started simply enough with the need to measure the couplings (and masses) that appear in our theory and to express physical predictions in terms of these measurements. However, it then got a bit more subtle. Some of the effects that we calculate (such as the unknown physics of very high energies) get lost because they become part of the coupling constants and masses that we have measured. This has the wonderful effect that the apparent divergences found in perturbation theory also disappear from physical predictions. It also leads to important techniques such as the renormalization group and effective field theory.

CHAPTER 8

Path integrals

Path integral methods are an alternate way to define quantum mechanics and Quantum Field Theory. The techniques feel somewhat scary and unnatural at first: how are we supposed to integrate over all possible field configurations? However, once you get used to the methods, they form a much simpler starting point for quantum physics. There are no operators, just functions. The classical limit makes sense. The techniques are more powerful, so that gauge theories and gravity are best quantized this way. They lend themselves to concrete numerical simulations and have a remarkable connection to Statistical Mechanics. In relativistic theories they are more manifestly covariant, as the starting point uses the Lagrangian rather than the Hamiltonian.

However, it is also true that the notion of quanta is not as readily defined. For this, canonical quantization is best. Also, we still need to impose the Born rule that Probability $= |\text{Amplitude}|^2$ when the path integral gives us an amplitude. So we are still building from the insights developed in other formalisms of quantization.

Still, the path integral formalism is a good unifying framework for physics. It should be part of the tool kit of all physicists.

8.1 Path integrals in quantum mechanics

Most quantum mechanics textbooks now contain chapters introducing path integrals. We will briefly review the development here for completeness. We can motivate the idea that propagation is the sum over all paths by generalizing the situation of a screen with two slits, where we know that the wavefunction includes propagation through both slits. If there were more slits, or even an infinite number of slits, the wavefunction would sample all the other slits also. Because the location of the screen is also arbitrary, it is clear that the propagation samples all paths through all possible slits.

The actual derivation of the path integral is just the reformulation in mathematical terms of the reasoning above. Consider the one-dimensional nonrelativistic propagator defined by

$$\psi(x_f, t_f) = \int dx_i\, D(x_f, t_f; x_i, t_i)\, \psi(x_i, t_i) \tag{8.1}$$

with

$$D\left(x_f, t_f; x_i, t_i\right) = \left\langle x_f \left| e^{-iH(t_f - t_i)} \right| x_i \right\rangle \equiv \left\langle x_f, t_f | x_i, t_i \right\rangle . \tag{8.2}$$

If we break the time interval up into N steps (i.e., like N screens) with $\delta t = (t_f - t_i)/N$ and insert a complete set of states at each (i.e., like an infinite number of slits),

$$\mathbf{1} = \int_{-\infty}^{\infty} dx_n\, |x_n\rangle\, \langle x_n| , \tag{8.3}$$

then we end up with

$$D\left(x_f, t_f; x_i, t_i\right) = \int_{-\infty}^{\infty} dx_{N-1} \ldots \int_{-\infty}^{\infty} dx_1$$
$$\times \left\langle x_N \left| e^{-iH\,\delta t} \right| x_{N-1} \right\rangle \left\langle x_{N-1} \left| e^{-iH\,\delta t} \right| x_{N-2} \right\rangle \ldots \left\langle x_1 \left| e^{-iH\,\delta t} \right| x_0 \right\rangle , \tag{8.4}$$

where we denoted $x_0 \equiv x_i$ and $x_N \equiv x_f$. We now have, to first order in δt

$$\left\langle x_\ell \left| e^{-iH\delta t} \right| x_{\ell-1} \right\rangle = \left\langle x_\ell \left| e^{-i\delta t\left[\frac{\hat{p}^2}{2m} + V(x) \right]} \right| x_{\ell-1} \right\rangle \tag{8.5}$$
$$= e^{-iV(x_\ell)\,\delta t} \left\langle x_\ell \left| e^{-i\delta t \frac{\hat{p}^2}{2m}} \right| x_{\ell-1} \right\rangle + \mathcal{O}\left((\delta t)^2\right).$$

For the kinetic energy part of the Hamiltonian we can use a complete set of momentum eigenstates to write

$$\left\langle x_\ell \left| e^{-i\delta t \frac{\hat{p}^2}{2m}} \right| x_{\ell-1} \right\rangle = \lim_{\epsilon \to 0^+} \int_{-\infty}^{\infty} \frac{dp}{2\pi} \langle x_\ell | p \rangle e^{-i\delta t \frac{p^2}{2m} - \epsilon p^2} \langle p | x_{\ell-1} \rangle$$
$$= \lim_{\epsilon \to 0^+} \int_{-\infty}^{\infty} \frac{dp}{2\pi} e^{ip(x_\ell - x_{\ell-1}) - i\delta t \frac{p^2}{2m} - \epsilon p^2} \tag{8.6}$$
$$= \sqrt{\frac{m}{2\pi i\, \delta t}}\, e^{i\frac{m}{2\delta t}(x_\ell - x_{\ell-1})^2} ,$$

where ϵ is temporarily introduced as a regulator to ensure that the Gaussian integral with an imaginary argument is well defined. If we take the $N \to \infty$ limit of the result, we have

$$D\left(x_f, t_f; x_i, t_i\right) =$$

$$\lim_{N\to\infty} \left(\frac{m}{2\pi i \,\delta t}\right)^{\frac{N}{2}} \left[\prod_{\ell=1}^{N-1} \int_{-\infty}^{\infty} dx_\ell\right] e^{i\sum_{\ell=1}^{N}\left[\frac{m}{2}\frac{(x_\ell - x_{\ell-1})^2}{\delta t} - V(x_\ell)\delta t\right]}. \tag{8.7}$$

This tells us that the propagation is described by integrals over all possible paths. The weight turns out to be the action

$$S[x(t)] = \lim_{N\to\infty} \sum_{\ell=1}^{N} \delta t \left(\frac{m}{2}\frac{(x_\ell - x_{\ell-1})^2}{(\delta t)^2} - V(x_\ell)\right)$$

$$= \int dt \left(\frac{m}{2}\dot{x}^2 - V(x)\right). \tag{8.8}$$

To clean the notation a little up we can define

$$\int \mathcal{D}[x(t)] \equiv \lim_{N\to\infty} \left(\frac{m}{2\pi i \,\delta t}\right)^{N/2} \prod_{n=1}^{N-1} \int_{-\infty}^{\infty} dx_n, \tag{8.9}$$

which leads to

$$\boxed{D\left(x_f, t_f; x_i, t_i\right) = \int \mathcal{D}[x(t)] e^{i\int_{t_i}^{t_f} dt\, L(x(t), \dot{x}(t))}}. \tag{8.10}$$

This is the path integral.

Defining the path integral was the easy part. It is still not clear how to use it in a practical way. For the most part we will describe this once we have transitioned to field theory. But there is one aspect that can be easily described here. This involves obtaining the ground state properties. If we insert a complete set of energy eigenstates $\psi_n(x)$ into the definition of the propagator, equation (8.2), we readily find

$$D\left(x_f, t_f; x_i, t_i\right) = \sum_n \psi_n^*(x_f)\,\psi_n(x_i)\,e^{-iE_n(t_f - t_i)}. \tag{8.11}$$

The long time behavior of this object preferentially isolates the ground state. A heuristic argument is the following: we can always set the energy of the ground state to 0, and then as $t_f - t_i \to \infty$, the excited states oscillate infinitely and hence get washed out, while the ground state does not oscillate. To put this argument on more solid grounds, one needs a way to make the oscillatory sum better behaved. One way is to analytically continue to imaginary time, $t_f - t_i \to -i\tau$. Another way is to add a small imaginary part to the energy

$$D\left(x_f, t_f; x_i, t_i\right) \to \sum_n \psi_n^*(x_f)\,\psi_n(x_i)\,e^{-i(E_n - i\epsilon)(t_f - t_i)}, \tag{8.12}$$

with $\epsilon > 0$. Either way we see reliably that

$$D\left(x_f, t \to \infty; x_i, 0\right) \to \sum_n \psi_0^*(x_f)\,\psi_0(x_i)\,e^{-i(E_0 - i\epsilon)t}\,, \tag{8.13}$$

and we can recover ground state properties.

Within the quantum mechanical path integral context, we can make this modification by using the relationship between the Hamiltonian and the Lagrangian. The modification

$$D\left(x_f, t_f; x_i, t_i\right) \to \int \mathcal{D}[x(t)]\,e^{i\int_{t_i}^{t_f} dt[L(x,\dot{x}) + i\epsilon\,x^2]} \tag{8.14}$$

has the effect of both making the path integral itself better defined and also adding a negative imaginary part to the energy. Evaluated in the large time limit, this path integral then also isolates the ground state.

8.2 Path integrals for Quantum Field Theory

In chapter 2 we saw that one can arrive at a field theory by treating the field variables as if they were coordinates. This was an exact identification in the phonon example that also yielded the field commutation rules. This same identification works with path integrals. In the phonon example, the path integral would initially involve integrals over each of the coordinates,

$$\int \prod_i \mathcal{D}[\delta y_i(t)]\,e^{iS}\,. \tag{8.15}$$

After transition to continuum variables $\delta y_i(t) \sim \phi(t, x)$, this would read

$$N \int [d\phi(t, x)]\,e^{iS}\,, \tag{8.16}$$

where N is a constant, which in the end will not be important. The notation $[d\phi(t, x)]$ will be what we use for the integration over fields at all spacetime points, analogous to the notation $\mathcal{D}[x(t)]$ for the integration over all paths. The integration over the fields will take some effort to get used to. It can be better defined by making spacetime discrete and then taking the number of points to infinity. What this path integral implies is then

$$D \sim N \prod_{i,j,k,\ell} \int_{-\infty}^{\infty} [d\phi\left(x_i, y_j, z_k, t_\ell\right)]\,e^{iS[\phi(x_i,y_k,z_k,t_\ell), \partial_\mu \phi(x_i,y_j,z_k,t_\ell)]}\,, \tag{8.17}$$

which we will abbreviate as

$$\int [d\phi(x)] \, e^{iS[\phi(x),\partial_\mu\phi(x)]} \, . \tag{8.18}$$

In these expressions, the action can be evaluated with initial starting configurations at some initial time and also the final configurations at a final time.

But what is this object that we have obtained in the path integral, and how can we evaluate it? This can become clearer by an explicit calculation. The first step is to extend the time integration to infinity, which when done carefully will pick out the ground state much like the quantum mechanical example of section 8.2. Because the ground state in Quantum Field Theory is the vacuum, the result will be vacuum matrix elements. In addition, using a scalar field as an example, we will add a *source term* $J(x)\,\phi(x)$ to the Lagrangian. The source $J(x)$ is an arbitrary function of spacetime that acts as a probe as it excites the field ϕ. The equation of motion is now

$$(\Box + m^2)\,\phi(x) = J(x)\,, \tag{8.19}$$

which has the general solution

$$\phi(x) = \int d^4x' \, G(x,\, x')\, J(x') \tag{8.20}$$

in terms of the Green's function $G(x,\, x')$. This will allow us to probe the path integral in various ways by looking at the ways the field $\phi(x)$ responds to the source $J(x)$. For a start we will do this with the free field theory and then return to the interacting theory in section 8.3.

Specifically, we define a *generating functional*

$$Z_0[J] = N \int [d\phi(x)] \, e^{i \int d^4x \left(\frac{1}{2}\partial_\mu\phi\partial^\mu\phi - \frac{1}{2}m^2\phi^2 + J\phi\right)} \, , \tag{8.21}$$

where the integration is now taken over all time (and space). The term "functional" implies that Z_0 depends on the whole functional form of $J(x)$. The subscript 0 on $Z_0[J]$ indicates that this is only the free field theory, without interactions. The exponential is a quadratic functional of the field $\phi(x)$. However, the exponent is purely imaginary and to ensure that the path integral is convergent, we add an infinitesimal factor $+i\epsilon\phi^2$ to the Lagrangian (with ϵ a positive infinitesimal). Combined with the overall factor of i in front of the action this adds a damping factor $-\epsilon\phi^2$ to the exponent.[1] As typical with Gaussian integrals, this path integral can be solved by a "completing the square" method. Note that we can use integration by parts to rewrite the exponent as

$$\int d^4x \left(\frac{1}{2}\partial_\mu\phi\partial^\mu\phi - \frac{1}{2}(m^2 - i\epsilon)\phi^2\right) = -\int d^4x \, \frac{1}{2}\phi\left(\Box + m^2 - i\epsilon\right)\phi\,. \tag{8.22}$$

---

[1] Equivalently one could analytically continue to imaginary time. We will see that this choice for the convergence factor is compatible with the usual Wick rotation performed to evaluate loop integrals, as discussed in the Appendix.

Using this we can define a new field

$$\phi'(x) = \phi(x) + \int d^4y \, D_F(x - y) \, J(y), \qquad (8.23)$$

where the function $D_F(x)$ is defined by requiring

$$\left(\Box_x + m^2 - i\epsilon\right) D_F(x - y) = -\delta^{(4)}(x - y), \qquad (8.24)$$

and is immediately identified as the Feynman propagator

$$iD_F(x - y) = i \int \frac{d^4k}{(2\pi)^4} \frac{e^{-ik\cdot(x-y)}}{k^2 - m^2 + i\epsilon}. \qquad (8.25)$$

This redefinition is useful as it separates the exponential into two parts

$$e^{-\frac{i}{2} \int d^4x \, \phi'(\Box + m^2 - i\epsilon)\phi'} \times e^{-\frac{i}{2} \int d^4x \, d^4y \, J(x) \, D_F(x-y) \, J(y)}. \qquad (8.26)$$

The path integral can now be done because the integration over ϕ' is the same as that over ϕ,

$$\int [d\phi'] = \int [d\phi]. \qquad (8.27)$$

Using this procedure, we obtain

$$Z_0[J] = Z_0[0] \, \exp\left\{-\frac{1}{2} \int d^4x \, d^4y \, J(x) \, iD_F(x - y) \, J(y)\right\}. \qquad (8.28)$$

Even though path integration can seem scary, this calculation was actually not that hard and the result is an understandable integral over functions.

To use this result, we need to introduce the technique of *functional differentiation*. The result equation (8.28) is a functional of the source $J(x)$, in that it depends on that function at all points of spacetime. We can learn about its behavior at different points by defining the differentiation rule

$$\frac{\delta J(y)}{\delta J(x)} = \delta^{(4)}(x - y) \implies \frac{\delta\left(\int d^4y \, J(y) f(y)\right)}{\delta J(x)} = f(x). \qquad (8.29)$$

Using this, we can readily find that when we first differentiate and then set J to 0 we will obtain a key result,

$$-\frac{1}{Z_0[0]} \frac{\delta^2 Z_0[J]}{\delta J(x_1) \, \delta J(x_2)}\bigg|_{J=0} = iD_F(x_1 - x_2)$$

$$= \langle 0| T \left(\phi(x_1) \, \phi(x_2)\right)|0\rangle, \qquad (8.30)$$

where the second line is the original definition of the Feynman propagator.[2] As expected, this is a vacuum matrix element, but in this case it is a very important one. Moreover, if we return to the original definition of the generating functional equation (8.21) and apply functional differentiation to that form, we see the identification

$$\langle 0|T\left(\phi(x_1)\,\phi(x_2)\right)|0\rangle = \frac{\int [d\phi(x)]\,\phi(x_1)\,\phi(x_2)\,e^{iS[\phi(x),\partial_\mu\phi(x)]}}{\int [d\phi(x)]\,e^{iS[\phi(x),\partial_\mu\phi(x)]}}\,. \tag{8.31}$$

This form of identification can be extended to other vacuum matrix elements. If we define the n-point vacuum element

$$G^{(n)}(x_1,\ldots,x_n) = \langle 0\,|T\left(\phi(x_1)\ldots\phi(x_n)\right)|\,0\rangle\,, \tag{8.32}$$

then its path integral version reads

$$\boxed{G^{(n)}(x_1,\ldots,x_n) = \frac{\int [d\phi(x)]\,\phi(x_1)\ldots\phi(x_n)\,e^{iS[\phi(x),\partial_\mu\phi(x)]}}{\int [d\phi(x)]\,e^{iS[\phi(x),\partial_\mu\phi(x)]}}} \tag{8.33}$$

and it can be calculated from the generating functional via differentiation

$$G^{(n)}(x_1,\ldots,x_n) = (-i)^n\,\frac{1}{Z[0]}\,\frac{\delta^n}{\delta J(x_1)\ldots\delta J(x_n)}Z[J]\bigg|_{J=0}\,. \tag{8.34}$$

These relations continue to hold in an interacting theory, which is why we have left off the subscript 0 on Z in equation (8.34). We have the tools at present to calculate these in the free field theory, where we find

$$G^{(2)}(x_1,x_2) = \frac{(-i)^2}{Z_0[0]}\,\frac{\delta^2}{\delta J(x_1)\,\delta J(x_2)}Z_0[J]\bigg|_{J=0} = iD_F(x_1-x_2)$$

$$G^{(4)}(x_1,x_2,x_3,x_4) = \frac{(-i)^4}{Z_0[0]}\,\frac{\delta^4}{\delta J(x_1)\,\delta J(x_2)\,\delta J(x_3)\,\delta J(x_4)}Z_0[J]\bigg|_{J=0} \tag{8.35}$$

$$= G^{(2)}(x_1,x_2)\,G^{(2)}(x_3,x_4) + G^{(2)}(x_1,x_3)\,G^{(2)}(x_2,x_4)$$

$$+ G^{(2)}(x_1,x_4)\,G^{(2)}(x_2,x_3)\,.$$

---

[2]Note that if we did not set $J(x) = 0$ after taking the functional derivatives, we would have obtained the Feynman propagator in the presence of a source.

The latter is just the statement of Wicks's theorem, which we can write in a diagrammatic form as

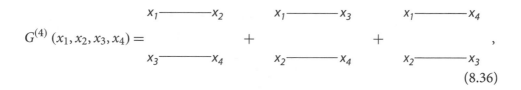

$$G^{(4)}(x_1, x_2, x_3, x_4) = \qquad + \qquad + \qquad , \tag{8.36}$$

where the symbol $x_1 \text{———} x_2$ denotes the propagator $iD_F(x_1 - x_2)$.

Next we turn to the addition of interactions.

8.3 The generating functional—Feynman rules again

The technique introduced in section 8.2 allows us to derive the Feynman rules in a different way. To set up for this task, let us consider a theory with interactions, which we will for the moment just refer to as $\mathcal{L}_I(\phi)$. The full path integral is then

$$Z[J] = N \int [d\phi(x)] e^{i \int d^4x [\mathcal{L}_0(\phi) + \mathcal{L}_I(\phi) + J(x)\phi(x)]}, \tag{8.37}$$

with $\mathcal{L}_0$ being the free field Lagrangian. When including the interactions, this cannot be solved in general. However it can be reformulated in a way that is amenable to a perturbative expansion.

Let us show how the generating functional can be rewritten using functional differentiation. The interaction term can be changed from $\mathcal{L}_I(\phi(x))$ to $\mathcal{L}_I(-i\delta/\delta J(x))$ and then taken outside of the path integral. Specifically we can see that

$$Z[J] = N \int [d\phi(x)] e^{i \int d^4y \, \mathcal{L}_I(\phi(y))} e^{i \int d^4x (\mathcal{L}_0(\phi) + J(x)\phi(x))}$$

$$= e^{i \int d^4y \, \mathcal{L}_I(-i\delta/\delta J(y))} N \int [d\phi(x)] e^{i \int d^4x (\mathcal{L}_0(\phi) + J(x)\phi(x))}$$

$$= e^{i \int d^4y \mathcal{L}_I(-i\delta/\delta J(y))} Z_0[J]. \tag{8.38}$$

This trick has converted an impossible path integral into an exercise in differentiation. This is still not completely solved because the exponential is an infinite series. But when we expand the result in powers of the coupling constant, we recover perturbation theory and the Feynman rules.

We can see how this works by examining a specific example in detail. Using the $\lambda\phi^4$ interaction and expanding the exponential in powers of λ, we have

$$Z[J] = Z_0[J] + Z_1[J] + Z_2[J] + \dots . \tag{8.39}$$

The first-order term is

$$Z_1[J] = -i\frac{\lambda}{4!} \int d^4y \, (-i)^4 \frac{\delta^4 Z_0[J]}{\delta J(y)^4} \, . \tag{8.40}$$

One can see that there are many ways for the derivative to act on $Z_0[J]$. It is now a good point to introduce the following graphic representations:

$$x \text{———} y \;=\; iD_F(x-y) \, ,$$

$$\bullet \text{———} y \;=\; \int d^4x \, J(x) \, iD_F(x-y) \, ,$$

$$\bullet \text{———} \bullet \;=\; \int d^4x \, d^4y \, J(x) \, iD_F(x-y) \, J(y) \, , \tag{8.41}$$

so that, for instance, $Z_0[J] = Z_0[0] \, \exp\{-\tfrac{1}{2}\bullet\text{—}\bullet\}$.

Note in particular that the diagrams including bullets vanish when the source vanishes:

$$\bullet \text{———} y \, \Big|_{J=0} \;=\; \bullet\text{———}\bullet \, \Big|_{J=0} = 0 \, . \tag{8.42}$$

Let us start by taking one derivative of $Z[J]$. We get

$$-i\frac{\delta Z_0[J]}{\delta J(x)} = -\int d^4z \, D_F(x-z) \, J(z) \, Z_0[J] = i\left(\bullet\text{———}\,x\right) e^{-\tfrac{1}{2}\bullet\text{—}\bullet} Z_0[0] \, . \tag{8.43}$$

When we take a second derivative of $Z[J]$ with respect to $J(y)$, we could act on the prefactor, in which case we end up with $D_F(x-y) \, Z_0[J]$, or we could act on $Z_0[J]$ again, bringing down another factor from the exponent:

$$(-i)^2 \frac{\delta^2 Z_0[J]}{\delta J(y) \, \delta J(x)} = \left[x \text{———} y - \left(\bullet\text{———}\,y \times \bullet\text{———}\,x\right)\right]$$

$$\times e^{-\tfrac{1}{2}\bullet\text{—}\bullet} Z_0[0] \, , \tag{8.44}$$

so that if we set $y = x$ and $J = 0$, we are left with the one-loop diagram

$$(-i)^2 \frac{\delta^2 Z_0[J]}{\delta J(x)^2}\Big|_{J=0} = \left(\bigcirc x - \text{—}\!\!\!> x\right) \times e^{-\tfrac{1}{2}\bullet\text{—}\bullet} Z_0[0]\Big|_{J=0}$$

$$= \left(\bigcirc x\right) Z_0[0] \, . \tag{8.45}$$

We see here how some of the elements of the Feynman rules start to emerge in a path integral framework.

Proceeding with this set of techniques, we obtain

$$Z_1[J] = -i\frac{\lambda}{4!} \int d^4y \, (-i)^4 \frac{\delta^4 Z_0[J]}{\delta J(y)^4}$$

$$= -i\frac{\lambda}{4!} \int d^4y \left(3 \;\; \bigotimes^y \;\; - \; 6 \; \bullet\!\!-\!\!\underset{y}{\overset{\bigcirc}{}}\!\!-\!\!\bullet \;\; + \;\; \bigtimes\!\!\!^y \right) Z_0[J]. \qquad (8.46)$$

For instance, the term with four external sources can be written explicitly as

$$Z_1[J] = -i\frac{\lambda}{4!} \int d^4y \int d^4z_1 \, J(z_1) \, D_F(z_1 - y) \int d^4z_2 \, J(z_2) \, D_F(z_2 - y)$$

$$\times \int d^4z_3 \, J(z_3) \, D_F(z_3 - y) \int d^4z_4 \, J(z_4) \, D_F(z_4 - y) \, Z_0[J] + \dots. \qquad (8.47)$$

As a first exercise let us calculate the two-point Green's function. We have already seen that in the free field theory this is the Feynman propagator, as in equation (8.35). Now we want to calculate this in the interacting theory. To $O(\lambda)$, this is given by

$$G^{(2)}(x_1, x_2) = \frac{(-i)^2}{Z_0[0] + Z_1[0]} \frac{\delta^2 \, (Z_0[J] + Z_1[J])}{\delta J(x_1) \, \delta J(x_2)} \bigg|_{J=0} + O(\lambda^2), \qquad (8.48)$$

where the denominator contains

$$Z_0[0] + Z_1[0] = \left(1 + i\frac{\lambda}{4!} \int d^4y \, 3 \;\; \bigotimes^y \right) Z_0[0] \qquad (8.49)$$

and whereas the differentiation of the numerator can be written as

$$(-i)^2 \frac{\delta^2}{\delta J(x_1) \, \delta J(x_2)} \left[1 - i\frac{\lambda}{4!} \int d^4y \left(3 \;\; \bigotimes^y \;\; - \; 6 \; \bullet\!\!-\!\!\underset{y}{\overset{\bigcirc}{}}\!\!-\!\!\bullet \;\; + \;\; \bigtimes\!\!\!^y \right) \right] Z_0[J].$$
$$(8.50)$$

Here the differentiation will act on $Z_0[J]$ and on the prefactor. The differentiation of $Z_0[J]$ gives the free field propagator, which is multiplied by the same factor that appears on the right-hand side of equation (8.49) and is thus canceled by the denominator. (Thus, as we saw in section 5.5, the disconnected diagrams cancel out.) The differentiation of the prefactor yields a new term, which is the self-energy diagram. Overall, the result can be written as

$$G^{(2)}(x_1, x_2) = x_1 \underline{\quad\quad} x_2 + i\frac{\lambda}{2} \int d^4y \left[x_1 \underset{y}{\underline{\quad \varphi \quad}} x_2 \right] + O(\lambda^2)$$

$$= iD_F(x_1 - x_2) + i\frac{\lambda}{2} \int d^4y\, iD_F(x_1 - y)\, iD_F(x_2 - y)\, iD_F(y - y) + O(\lambda^2)$$

$$(8.51)$$

or, taking a Fourier transform,

$$\frac{i}{p^2 - m^2 + i\epsilon} + \frac{i}{p^2 - m^2 + i\epsilon} \int \frac{d^4q}{(2\pi)^4} \frac{-\lambda/2}{q^2 - m^2 + i\epsilon} \frac{i}{p^2 - m^2 + i\epsilon}, \qquad (8.52)$$

which is the same result that one finds with the usual Feynman rules.

In a similar fashion we can derive the first-order contribution to the four-point Green's function,

$$G^{(4)}(x_1, \ldots, x_4) = (-i)^4 \frac{1}{Z[0]} \frac{\delta^4}{\delta J(x_1) \ldots \delta J(x_4)} Z[J]\bigg|_{J=0}. \qquad (8.53)$$

When acting on the term in Z_1 written in equation (8.47), all four derivatives must act on the various prefactors of $J(z_i)$ in order to remove them or else the result will vanish when we set $J = 0$. The result for this term then is

$$G^{(4)}(x_1, x_2, x_3, x_4) = -i\lambda \int d^4y \left[\begin{array}{c} x_1 \diagdown \quad \diagup x_2 \\ \times \\ x_3 \diagup y \diagdown x_4 \end{array} \right]$$

$$= -i\lambda \int d^4y\, iD_F(x_1 - y)\, iD_F(x_2 - y)\, iD_F(x_3 - y)\, iD_F(x_4 - y). \qquad (8.54)$$

Here we see again the start of the Feynman rules—the $-i\lambda$ for the vertex. Also included are the external propagators, giving the propagation into the vertex. However, there are other diagrams in the result that do not contribute to a scattering amplitude. The free field four-point function of equation (8.35), obtained by differentiating $Z_0[J]$ four times, does not contribute to scattering because, when Fourier transformed to momentum space, momentum conservation would require the initial and final momenta to be unchanged. A related set of diagrams, given diagrammatically by

$$\sim -i\lambda \int d^4y \left[\begin{array}{c} x_1 \underline{\quad\quad} x_2 \\ x_3 \underset{\bigcirc}{\underline{\quad y \quad}} x_4 \end{array} \right], \qquad (8.55)$$

is obtained by differentiating $Z_0[J]$ twice and $Z_1[J]$ twice and represents just the renormalization of the lowest order propagator and does not contribute for the same reason. Finally, disconnected diagrams cancel because in the end we divide by $Z[0]$. At the end, the only diagram that gives a net contribution to the scattering interaction is given by equation (8.54).

The interpretation illustrated in this example is that the source $J(x)$ probes the vacuum state by exciting a field. The interaction of the fields are represented in coordinate space by the Green's functions. We see the ingredients of the Feynman rules with the calculation of the n-point functions. The final question is: How do we go from those n-point functions in coordinate space to scattering amplitude for momentum eigenstates?

The exact relationship is given by the *LSZ reduction formula* (where LSZ refers to the authors Lehmann, Syzmanzik, and Zimmermann). For the four-point scattering amplitude, this reads

$$\langle p_1, p_2 | S | q_1, q_2 \rangle = \int \prod_{i=1,2} \left\{ d^4 x_i \, i \, \frac{e^{-iq_i z_i}}{\sqrt{Z}} \left(\Box_{x_i} + m^2 \right) \right\}$$

$$\times \prod_{j=1,2} \left\{ d^4 y_j \, i \, \frac{e^{ip_j y_j}}{\sqrt{Z}} \left(\Box_{y_j} + m^2 \right) \right\} \langle 0 | T \left(\varphi (x_1) \, \phi (x_2) \, \phi (y_1) \, \phi (y_2) \right) | 0 \rangle, \quad (8.56)$$

where Z is the wavefunction renormalization constant (which is unity at the level that we are presently calculating). The $\Box + m^2$ operators acting on the propagators in the Green's functions yield Dirac delta functions that are removed by the integration over position. In essence this tells us to get the scattering amplitude by dropping the external propagators and Fourier transforming the remainder. That leaves behind the momentum-space amplitude, which we have calculated by different means earlier in the book. This procedure also holds for scattering amplitudes with more particles. The generating functional contains *everything*—all of the Green's functions and scattering amplitudes of the theory.

This effort can be carried out to higher orders in the coupling. The principles are the same. Connected diagrams contribute to the scattering amplitude and give the usual results of the Feynman rules. Self-energy diagrams are accounted for by the renormalization procedure. Disconnected diagrams drop out because we divide by $Z[0]$.

The path integral treatment has avoided the construction of a field *operator*, instead it deals with fields as ordinary functions. But we have originally used the operator construction to define the states of the theory, that is, its quanta. How does the idea of quanta arise in a path integral context? If you review what we have done, you can see the existence of quanta in the poles of the propagator. In the original calculation of the propagator in section 4.6, the creation operators were used to give us the initial form of the propagator, describing the evolution of one quantum with $E = \hbar \omega = \sqrt{\mathbf{p}^2 + m^2}$. We then used complex integration to convert this into the usual form of the Feynman propagator, but it was the location of the poles that allowed this to be done. In the path integral context, we have used the source J to excite the field and have found the same propagator. The interpretation here is that the source excites the quanta of the field, which then scatter and decay. That is what the calculation in this section shows. So looking at the poles in the propagator identifies the particles.

While our exploration of path integrals has focused on obtaining the perturbative Feynman rules, the subject can lead to an even richer exploration of Quantum Field Theory. Much of this is beyond the scope of our introductory book. But the prime takeaway message is that one can define all of Quantum Field Theory starting from the path integral, which then provides an alternate entrance to quantum theory when compared to canonical quantization.

8.4 Connection to statistical physics

We started the discussion of path integrals with the nonrelativistic propagator

$$D\left(x_f, t_f; x_i, t_i\right) = \left\langle x_f \left| e^{-iH\left(t_f - t_i\right)} \right| x_i \right\rangle \tag{8.57}$$

and wrote a path integral for this object. As a mathematical function, we can play games with this. For instance, we can continue this object to imaginary time, $t_f - t_i \to -i(\tau_f - \tau_i)$, which gives the change

$$e^{-iHt} \to e^{-H\tau} . \tag{8.58}$$

By making the motivated choice of $\tau_i = 0$ and $\tau_f = \beta$, we obtain

$$\left\langle x_f \left| e^{-\beta H} \right| x_i \right\rangle . \tag{8.59}$$

The notation has been chosen so that $e^{-\beta H}$ stands out immediately as the density matrix for a system at temperature T if β is identified with $1/k_B T$, where k_B is the Boltzmann constant. If we wish to obtain the coordinate space representation of the partition function, we just need to take the trace,

$$\mathcal{Z} = \mathrm{Tr}\left\{e^{-\beta H}\right\} = \int dx \langle x \left| e^{-\beta H} \right| x \rangle . \tag{8.60}$$

To do this we have to identify $x_f = x_i$ or $x(\tau_i + \beta) = x(\tau_i)$, which are periodic boundary conditions.

So all the path integral techniques we have seen in this chapter can also be used to compute partition functions! The time development is now over a finite time and the action is

$$iS = i \int_{t_i}^{t_f} dt \left[\frac{m}{2} \left(\frac{dx}{dt}\right)^2 - V(x) \right] = -\int_0^\beta d\tau \left[\frac{m}{2} \left(\frac{dx}{d\tau}\right)^2 + V(x) \right], \tag{8.61}$$

which now includes the Hamiltonian. We thus obtain

$$\mathcal{Z} = N \int \mathcal{D}[x(t)] \, e^{-\int_0^\beta d\tau \left[\frac{1}{2} m \left(\frac{dx}{d\tau}\right)^2 + V(x)\right]} . \tag{8.62}$$

Here the periodicity condition $x(\tau_i + \beta) = x(\tau_i)$ applies for any particular end-points, but the path integral sums over all possible endpoints.

This connection continues in Quantum Field Theory. We analytically continue $t \to -i\tau$, take a finite time interval $0 \le \tau \le \beta$, and transition $x_i(\tau) \to \phi(\mathbf{x}, \tau)$ with periodic boundary conditions

$$\phi(\mathbf{x}, \beta) = \phi(\mathbf{x}, 0) . \tag{8.63}$$

With this change we have

$$\mathcal{L} \to -\frac{1}{2} \left(\frac{d\phi}{d\tau}\right)^2 - \frac{1}{2} (\nabla \phi)^2 - \frac{1}{2} m^2 \phi^2 - V(\phi) \equiv -\mathcal{L}_E , \tag{8.64}$$

which again looks similar to a Hamiltonian. The partition function can be then be written as

$$\mathcal{Z}[J] = \int [d\phi] \, e^{-\int_0^\beta d\tau \, d^3x \, (\mathcal{L}_E - J\phi)} . \tag{8.65}$$

This is a striking result.

Note also the physical interpretation that as the temperature goes to 0, $\beta \to \infty$, the thermal system ends up in the ground state. We previously defined the field-theoretical generating function in Minkowski space by using a regulator as in equations (8.12) and (8.22) to isolate the ground state. However, this indicates an alternative way to define $Z[J]$ by a Euclidean rotation $t_f - t_i \to -i\tau$, with the limit $\tau \to \infty$. If the ground state functional is to be used in Minkowski space processes, the Euclidean functional integral needs to be continued back to Minkowski space. Doing this carefully requires the introduction of the $i\epsilon$ factors into the propagators that we have discussed throughout this book.

The techniques of perturbative quantum field theory can be applied to the evaluation of this partition function. A full exploration is beyond the scope of this book. However, the basic point is that there are Feynman rules, as in regular Quantum Field Theory, and we can define a perturbative expansion. The most important difference is the form of the propagator. We wish to solve

$$\left(-\frac{\partial^2}{\partial \tau_x^2} - \nabla_x^2 + m^2\right) D_\beta(x - y) = \delta(\tau_x - \tau_y) \, \delta^{(3)}(\mathbf{x} - \mathbf{y}) . \tag{8.66}$$

Because of the periodic boundary conditions, the solution will involve a Fourier sum rather than a Fourier integral. This requires the periodic factor

$$e^{i\omega_n \tau} \quad \text{with} \quad \omega_n = \frac{2\pi n}{\beta} . \tag{8.67}$$

The solution with the correct normalization is

$$D_\beta(x - y) = T \sum_n \int \frac{d^3k}{(2\pi)^3} \frac{e^{i\omega_n \tau} e^{i\mathbf{k}\cdot(\mathbf{x}-\mathbf{y})}}{\omega_n^2 + \mathbf{k}^2 + m^2}. \tag{8.68}$$

The Feynman rules then proceed as defined in Euclidean space, but with the modifications

$$\frac{1}{k_E^2 + m^2} \to \frac{1}{\omega_n^2 + \mathbf{k}^2 + m^2}$$

$$\int \frac{d^4 k_E}{(2\pi)^4} \to T \sum_n \int \frac{d^3 k}{(2\pi)^3}. \tag{8.69}$$

Note that a careful treatment indicates that $T \sum_n \to \int dk_0/(2\pi)$ as $T \to 0$, so that the usual continuum limit is obtained.

It is a remarkable correspondence that four-dimensional Euclidean Quantum Field Theory on an interval $0 \leq \tau \leq \beta$ with periodic boundary conditions is equivalent to quantum Statistical Mechanics in three dimensions.

Chapter summary: It is fascinating that there are two completely different ways to define Quantum Field Theory. We started the book with canonical quantization because this builds most directly from the intuition we develop in usual Quantum Mechanics treatments. But in this chapter we have reached the same endpoint by starting from path integrals. Wow!

CHAPTER 9

A short guide to the rest of the story

At this point in the book you have seen the main conceptual and calculational foundations of Quantum Field Theory. This should give you an understanding of how the subject operates and the ability to use field theory language. There is more to the subject. This chapter discusses things we feel may be useful as you continue your journey in understanding Quantum Field Theory.

9.1 Quantizing other fields

In this book we have mostly focused on the real scalar field that associates to each point in spacetime a single Hermitian operator, $\phi(t, \mathbf{x}) = \phi^\dagger(t, \mathbf{x})$. But we also have seen the complex scalar field that associates to each point in spacetime a non-Hermitian operator, $\phi(t, \mathbf{x}) \neq \phi^\dagger(t, \mathbf{x})$, and that is equivalent to *two* Hermitian operators, through the identification $\phi(t, \mathbf{x}) = \frac{1}{\sqrt{2}} [\phi_1(t, \mathbf{x}) + i\phi_2(t, \mathbf{x})]$. And we have discussed the photon field, $A_\mu(t, \mathbf{x})$, that is described by *four* Hermitian operators, $A_0(t, \mathbf{x})$, $A_1(t, \mathbf{x})$, $A_2(t, \mathbf{x})$, and $A_3(t, \mathbf{x})$ (even if some of them are not physical due to the subtleties of gauge invariance).

What is the criterion that makes us declare that two (or four) Hermitian operators that are functions of space and time are part of the same field? The answer is based on symmetries. If there is some symmetry according to which various operators transform into each other, then we tend to say that those operators are components of the same field.[1] For instance, the complex scalar Lagrangian has a symmetry $\phi \mapsto e^{i\alpha} \phi$, which is equivalent to $\phi_1 \mapsto \phi_1 \cos\alpha - \phi_2 \sin\alpha$, and $\phi_2 \mapsto \phi_1 \sin\alpha + \phi_2 \cos\alpha$, that shows how ϕ_1 and ϕ_2 mix under the symmetry.

In the case of photons, the symmetry that mixes the four components of A^μ is the Lorentz symmetry: under a Lorentz transformation $x^\mu \mapsto (x')^\mu = \Lambda^\mu{}_\nu x^\nu$, the components of the photon field undergo the same transformation that the coordinates do: $A^\mu \mapsto (A')^\mu = \Lambda^\mu{}_\nu A^\nu$. Are there other manifestations (or, using the correct mathematical term, *representations*) of the Lorentz symmetry that can act on fields?

---

[1] As you can guess, this is not a rigorous criterion, and there are situations where, even if two objects mix into each other under some symmetry, it can be convenient to use a notation where they are treated as independent fields.

The answer is yes, and a very important one is the so-called *spinor* representation of the Lorentz group.

9.1.1 THE DIRAC FIELD

The goal here is to describe how Quantum Field Theory treats fermions. These fields are called *(Dirac) spinors* or *Dirac fields*. You can read about the spinor representations of Lorentz group in books on group theory. Here we will focus on the physics of spinor fields without paying too much attention to their group-theoretical properties. They are described, in four spacetime dimensions by *four* complex fields that are usually presented in the form of a column vector,

$$\psi(t, \mathbf{x}) = \begin{pmatrix} \psi_1(t, \mathbf{x}) \\ \psi_2(t, \mathbf{x}) \\ \psi_3(t, \mathbf{x}) \\ \psi_4(t, \mathbf{x}) \end{pmatrix}, \tag{9.1}$$

and that satisfy the *Dirac equation*

$$\left(i\gamma^\mu \partial_\mu - m\right) \psi = 0, \tag{9.2}$$

where m is a real constant, which is understood to multiply the 4×4 identity matrix, and where the four γ^μs are 4×4 matrices that must satisfy the condition

$$\gamma^\mu \cdot \gamma^\nu + \gamma^\nu \cdot \gamma^\mu = 2g^{\mu\nu}, \tag{9.3}$$

where $g^{\mu\nu}$ is the Lorentz metric tensor and where the right-hand side is understood to be multiplied by the 4×4 identity matrix. One can choose any form for the gamma-matrices, as long as they satisfy equation (9.3). Here we will choose the so-called Dirac representation,

$$\gamma^0 = \begin{pmatrix} 1_2 & 0_2 \\ 0_2 & -1_2 \end{pmatrix}, \qquad \gamma^i = \begin{pmatrix} 0_2 & \sigma^i \\ -\sigma^i & 0_2 \end{pmatrix}, \tag{9.4}$$

where 1_2 denotes the 2×2 identity matrix, 0_2 denotes the 2×2 zero matrix, and σ^i are the three Pauli matrices

$$\sigma^1 = \begin{pmatrix} 0 & 1 \\ 1 & 0 \end{pmatrix}, \quad \sigma^2 = \begin{pmatrix} 0 & -i \\ i & 0 \end{pmatrix}, \quad \sigma^3 = \begin{pmatrix} 1 & 0 \\ 0 & -1 \end{pmatrix}. \tag{9.5}$$

The Lagrangian density reads

$$\boxed{\mathcal{L} = \bar{\psi}\left(i\partial\!\!\!/ - m\right)\psi}, \tag{9.6}$$

where we have introduced the Dirac "slash" notation $\rlap{/}{X} \equiv \gamma^\mu X_\mu$, and where we have defined

$$\bar{\psi} \equiv \psi^\dagger \gamma^0 \,. \tag{9.7}$$

The Dirac equation can thus be trivially obtained by varying the action with respect to $\bar{\psi}$ at fixed ψ, or, with a little more work, by varying the Lagrangian with respect to ψ while keeping $\bar{\psi}$ fixed. The momentum conjugate to ψ is $\Pi_\psi = \frac{\partial \mathcal{L}}{\partial \dot{\psi}} = i\psi^\dagger$, so that the Hamiltonian is given by

$$H = \int d^3x \, \bar{\psi} \left(-i\gamma^i \partial_i + m \right) \psi \,. \tag{9.8}$$

We can look for solutions of the Dirac equation in the form of plane waves, where all the space dependence is in a factor $e^{-ip\cdot x}$, finding two general solutions with that x^μ-dependence as

$$\psi(t, \mathbf{x}) = e^{-ip\cdot x} u_\lambda(\mathbf{p}), \quad u_\lambda(\mathbf{p}) = \frac{\sqrt{E+m}}{\sqrt{2E}} \begin{pmatrix} \chi_\lambda(\hat{\mathbf{p}}) \\ \dfrac{\boldsymbol{\sigma} \cdot \mathbf{p}}{E+m} \cdot \chi_\lambda(\hat{\mathbf{p}}) \end{pmatrix}, \quad \lambda = \pm, \tag{9.9}$$

where $p^0 = E \equiv \sqrt{\mathbf{p}^2 + m^2}$, $\hat{\mathbf{p}} = \mathbf{p}/p$, and where $\chi_\pm(\hat{\mathbf{p}})$ are any two linearly independent two-dimensional vectors. For what follows, it is convenient to choose them of the form

$$\chi_\lambda(\hat{\mathbf{p}}) = \frac{1 + \lambda\,\boldsymbol{\sigma} \cdot \hat{\mathbf{p}}}{\sqrt{2(1 + \hat{p}_3)}} \bar{\chi}_\lambda, \qquad \bar{\chi}_+ = \begin{pmatrix} 1 \\ 0 \end{pmatrix}, \quad \bar{\chi}_- = \begin{pmatrix} 0 \\ 1 \end{pmatrix} \tag{9.10}$$

that have the property $\boldsymbol{\sigma} \cdot \hat{\mathbf{p}}\, \chi_\lambda(\hat{\mathbf{p}}) = \lambda\, \chi_\lambda(\hat{\mathbf{p}})$, which we will find useful in the discussion surrounding equation (9.23).

Note that the overall coefficient in equation (9.9) is arbitrary, and we chose it to normalize $u_\lambda^\dagger(\mathbf{p}) \cdot u_{\lambda'}(\mathbf{p}) = \delta_{\lambda\lambda'}$, which implies $\bar{u}_\lambda(\mathbf{p}) \cdot u_{\lambda'}(\mathbf{p}) = \frac{m}{E}\delta_{\lambda\lambda'}$. Two more solutions can be found by looking for "negative frequency" modes, which are modes where the spacetime dependence is of the form $e^{+ip\cdot x}$, yielding the solutions

$$\psi(t, \mathbf{x}) = e^{ip\cdot x} v_\lambda(\mathbf{p}), \quad v_\lambda(\mathbf{p}) = \frac{\sqrt{E+m}}{\sqrt{2E}} \begin{pmatrix} \dfrac{\boldsymbol{\sigma} \cdot \mathbf{p}}{E+m} \cdot \chi_\lambda(\hat{\mathbf{p}}) \\ \chi_\lambda(\hat{\mathbf{p}}) \end{pmatrix}, \quad \lambda = \pm. \tag{9.11}$$

These solutions are also normalized to $v_\lambda^\dagger(\mathbf{p}) \cdot v_{\lambda'}(\mathbf{p}) = \delta_{\lambda\lambda'}$. Useful relations that can be obtained with a direct calculation are

$$\sum_{\lambda=\pm} u_\lambda(\mathbf{p})\, \bar{u}_\lambda(\mathbf{p}) = \frac{\rlap{/}{p} + m}{2E}, \qquad \sum_{\lambda=\pm} v_\lambda(\mathbf{p})\, \bar{v}_\lambda(\mathbf{p}) = \frac{\rlap{/}{p} - m}{2E}\,. \tag{9.12}$$

We can now quantize the Dirac field. Because this is a complex field, it is natural to decompose it in a similar manner as we would to decompose the complex scalar:

using two sets of creation/annihilation operators that, for the Dirac field, are usually denoted by $\hat{b}_{\mathbf{p},\lambda}^{(\dagger)}$ and $\hat{d}_{\mathbf{p},\lambda}^{(\dagger)}$. We decompose then the operator $\psi(\mathbf{x}, t)$ as

$$\psi(t, \mathbf{x}) = \sum_{\lambda=\pm} \int \frac{d^3p}{(2\pi)^3} [e^{-ip\cdot x} u_\lambda(\mathbf{p}) \hat{b}_{\mathbf{p},\lambda} + e^{ip\cdot x} v_\lambda(\mathbf{p}) \hat{d}_{\mathbf{p},\lambda}^\dagger]. \qquad (9.13)$$

The normalization of the coefficients of $\hat{b}_{\mathbf{p},\lambda}$ and $\hat{d}_{\mathbf{p},\lambda}^\dagger$ in this definition is chosen so that the Hamiltonian in equation (9.8) takes the suggestive form

$$H = \sum_{\lambda=\pm} \int \frac{d^3p}{(2\pi)^3} \sqrt{\mathbf{p}^2 + m^2} \, [\hat{b}_{\mathbf{p},\lambda}^\dagger \hat{b}_{\mathbf{p},\lambda} - \hat{d}_{\mathbf{p},\lambda} \hat{d}_{\mathbf{p},\lambda}^\dagger], \qquad (9.14)$$

which, based on our previous experience with the Hamiltonian of the complex scalar, suggests that $\hat{b}_{\mathbf{p},\lambda}$ might be the annihilation operator of a particle of kind λ and $\hat{d}_{\mathbf{p},\lambda}$ the annihilation operator of an antiparticle of kind λ. We will show in equation (9.21) that this is indeed the case, and in equation (9.23) we will also see what the quantum number λ corresponds to in physical terms.

You might have noticed, however, that the *sign* in front of $\hat{d}_{\mathbf{p},\lambda} \hat{d}_{\mathbf{p},\lambda}^\dagger$ in equation (9.14) is opposite to that found in the corresponding equation (3.66) for the Hamiltonian of the complex scalar! This is due to the fact that, unlike the complex scalars, the Dirac field describes *fermions*, which are fields where the usual commutation relations are replaced by *anticommutation* relations.

This means that the canonical quantization condition $[\phi(t, \mathbf{x}), \pi(t, \mathbf{y})] = i\delta^{(3)}(\mathbf{x} - \mathbf{y})$, which is valid for a scalar field $\phi(t, \mathbf{x})$ and its conjugate momentum $\pi(t, \mathbf{x})$, turns, for the Dirac field, into an anticommutation relation

$$\{\psi(t, \mathbf{x})_i, \pi_\psi(t, \mathbf{y})_j\} = \{\psi(t, \mathbf{x})_i, i\psi^\dagger(t, \mathbf{y})_j\} = i\delta_{ij}\delta^{(3)}(\mathbf{x} - \mathbf{y}), \qquad (9.15)$$

where we have defined the *anticommutator*

$$\{A, B\} \equiv AB + BA. \qquad (9.16)$$

In particular, the anticommutation relation in equation (9.15) implies the relations

$$\left\{\hat{b}_{\mathbf{p},\lambda}, \hat{b}_{\mathbf{p}',\lambda'}\right\} = \left\{\hat{b}_{\mathbf{p},\lambda}^\dagger, \hat{b}_{\mathbf{p}',\lambda'}^\dagger\right\} = 0, \qquad (9.17)$$

$$\left\{\hat{b}_{\mathbf{p},\lambda}, \hat{b}_{\mathbf{p}',\lambda'}^\dagger\right\} = (2\pi)^3 \delta^{(3)}(\mathbf{p} - \mathbf{p}') \delta_{\lambda\lambda'}, \qquad (9.18)$$

and analogous relations for the $\hat{d}_{\mathbf{p},\lambda}$ and $\hat{d}_{\mathbf{p},\lambda}^\dagger$ operators.

In particular, equation (9.17) implies that $(\hat{b}_{\mathbf{p},\lambda}^\dagger)^2 = (\hat{d}_{\mathbf{p},\lambda}^\dagger)^2 = 0$, which, applied to the vacuum $|0\rangle$, implies that it is impossible to construct a state that contains two Dirac fermions with same value of the parameter λ and same momentum! This is nothing but *Pauli's exclusion principle* in action: it is impossible for two fermions to be in the same state.

Once equation (9.18) is used, the Dirac Hamiltonian takes the form

$$H = \sum_{\lambda=\pm} \int \frac{d^3p}{(2\pi)^3} \sqrt{\mathbf{p}^2 + m^2} \, [\hat{b}^\dagger_{\mathbf{p},\lambda} \hat{b}_{\mathbf{p},\lambda} + \hat{d}^\dagger_{\mathbf{p},\lambda} \hat{d}_{\mathbf{p},\lambda}] + \text{(divergent) constant},$$

(9.19)

which agrees with the equivalent results found for complex scalars in chapter 3.

As is the case for complex scalars, the Dirac field also has a conserved current that takes the form

$$J^\mu(x) = \bar{\psi}(x)\, \gamma^\mu \psi(x), \qquad \partial_\mu J^\mu = 0,$$

(9.20)

and is associated to a charge

$$Q(t) = \int d^3x \, J^0(t, \mathbf{x}) = \sum_{\lambda=\pm} \int \frac{d^3p}{(2\pi)^3} \, [\hat{b}^\dagger_{\mathbf{p},\lambda} \hat{b}_{\mathbf{p},\lambda} + \hat{d}_{\mathbf{p},\lambda} \hat{d}^\dagger_{\mathbf{p},\lambda}]$$

$$= \sum_{\lambda=\pm} \int \frac{d^3p}{(2\pi)^3} \, [\hat{b}^\dagger_{\mathbf{p},\lambda} \hat{b}_{\mathbf{p},\lambda} - \hat{d}^\dagger_{\mathbf{p},\lambda} \hat{d}_{\mathbf{p},\lambda}] + \text{(divergent) constant}.$$

(9.21)

So we find that the operators $\hat{b}^\dagger_{\mathbf{p},\lambda}$ do indeed create particles, and the operators $\hat{d}^\dagger_{\mathbf{p},\lambda}$ create antiparticles.

To find out the meaning of the quantum number λ, we note that the i-th component of the spin operator, in the case of Dirac spinors, takes the form

$$\Sigma^i = \frac{i}{8}\epsilon_{ijk} [\gamma^j, \gamma^k] = \frac{1}{2} \begin{pmatrix} \sigma^i & 0_2 \\ 0_2 & \sigma^i \end{pmatrix}.$$

(9.22)

One can then define the *helicity* operator $\hat{h}$ as the projection of the spin along the direction of the momentum, $\hat{h} = \hat{\mathbf{p}} \cdot \mathbf{\Sigma}$. It is then possible to find that both $u_\lambda(\mathbf{p})$ and $v_\lambda(\mathbf{p})$ are eigenstates of $\hat{h}$ with eigenvalue $\lambda/2$:

$$\hat{h}\, u_\lambda(\mathbf{p}) = \frac{\lambda}{2} u_\lambda(\mathbf{p}),$$

$$\hat{h}\, v_\lambda(\mathbf{p}) = \frac{\lambda}{2} v_\lambda(\mathbf{p}).$$

(9.23)

This explains the quantum number λ: the operator $\hat{b}^\dagger_{\mathbf{p},\lambda}$ creates a particle of momentum $\mathbf{p}$ and helicity $\lambda/2$, whereas the operator $\hat{d}^\dagger_{\mathbf{p},\lambda}$ creates an *antiparticle* of momentum $\mathbf{p}$ and helicity $\lambda/2$.

Before computing the Feynman propagator for the Dirac fermions, it is worth mentioning that, the same way as complex scalars ϕ can be imposed on a reality condition $\phi^\dagger = \phi$, which implies that particles are equal to antiparticles, we can impose also a reality condition on spin-1/2 fields. This condition reads $\psi = -i\gamma^2 \psi^*$. Fermions that satisfy this condition are called *Majorana fermions* and carry only two degrees of freedom, one with helicity $+1/2$ and one with helicity $-1/2$.

Finally, the Feynman propagator is in this case a 4×4 matrix $S_F(x - x')$, whose elements are given by

$$i\left(S_F(x - x')\right)_{ab} = \langle 0|T\left(\psi(\mathbf{x}, t)_a \bar{\psi}(\mathbf{x}', t')_b\right)|0\rangle, \qquad (9.24)$$

where the T-product, given that we are dealing with anticommuting objects, is defined as

$$T\left(\psi(\mathbf{x}, t)_a \bar{\psi}(\mathbf{x}', t')_b\right) = \Theta(t - t')\,\psi(\mathbf{x}, t)_a \bar{\psi}(\mathbf{x}', t')_b$$
$$- \Theta(t' - t)\,\bar{\psi}(\mathbf{x}', t')_b\,\psi(\mathbf{x}, t)_a \qquad (9.25)$$

(note the relative minus sign).

It is easy to see that $S_F(x - x')$ is a Green's function for the Dirac equation

$$\left(i\slashed{\partial} - m\right)_{ab} i\,S_F(x - x')_{bc} = i\delta^{(4)}(x - x')\,\delta_{ac}, \qquad (9.26)$$

and we can compute

$$S_F(x) = \int \frac{d^4p}{(2\pi)^4} \frac{\slashed{p} + m}{p^2 - m^2 + i\epsilon}\,e^{-ip\cdot x} = \int \frac{d^4p}{(2\pi)^4} \frac{1}{\slashed{p} - m + i\epsilon}\,e^{-ip\cdot x}. \qquad (9.27)$$

9.1.2 GAUGE BOSONS

We briefly presented in section 5.7 some results for the photon. Photons are the simplest example of gauge bosons, which are fields associated to local symmetries. We will see that these fields carry a Lorentz index, A_μ^i, which will imply that they have spin equal to unity. But gauge bosons can come in a bigger family, hence the index i on A_μ^i. Gauge fields occur in many contexts. While a full treatment of such fields is more subtle than we can treat here, we will present an overview of the subject.

The archetype of a gauge field is the photon. It is associated with a symmetry of the Lagrangian under a local phase transformation, which is combined with a gauge transformation of the photon field, such that the action is invariant. For a complex scalar field $\phi = \frac{1}{\sqrt{2}}(\phi_1 + i\phi_2)$ this transformation was

$$\phi(x) \to e^{i\alpha(x)}\phi(x). \qquad (9.28)$$

This can be generalized if the symmetry is larger. For example, consider a field with two complex components

$$\phi(x) = \frac{1}{\sqrt{2}} \begin{pmatrix} \phi_1(x) + i\phi_2(x) \\ \phi_3(x) + i\phi_4(x) \end{pmatrix}. \tag{9.29}$$

If the mass and interaction terms are the same for all the components, then this field can have a global symmetry

$$\phi \to \phi' = U\phi, \tag{9.30}$$

where U is a unitary 2×2 matrix, that is, a matrix satisfying $U^\dagger U = 1$. With this condition we have $\phi'^\dagger \phi' = \phi^\dagger \phi$. We are familiar with constructing unitary 2×2 matrices from our work on the Pauli matrices in section 9.1.1. Because a unitary matrix can be written as the exponential of a Hermitian one, $U = \exp(iM)$, with $M^\dagger = M$, we have

$$U = e^{i\alpha} e^{i\boldsymbol{\beta} \cdot \boldsymbol{\tau}}. \tag{9.31}$$

Here $e^{i\alpha}$ is an overall phase β^i, $i = 1, 2, 3$ form a vector of constant parameters; and $\tau^i = \sigma^i$ are the three traceless Hermitian Pauli matrices. The renaming of $\sigma^i \to \tau^i$ is somewhat conventional to indicate that the present use has nothing to do with spin. We here focus only on the $\boldsymbol{\beta}$ terms as the effect of the overall phase α is the same as the one we saw in the case of the photon.

At this stage, we have identified a global symmetry (called $SU(2)$) for this field if the Lagrangian is constructed using invariant functions such as $\phi^\dagger \phi$. There is a way to make this into a local symmetry, with $\beta^i(x)$ depending on position, which will bring us the the new gauge bosons. In a local symmetry $\phi \to \phi' = U(x)\phi$, with

$$U(x) = e^{i\boldsymbol{\beta}(x) \cdot \boldsymbol{\tau}}. \tag{9.32}$$

This still leaves $\phi^\dagger \phi$ invariant but not $\partial_\mu \phi^\dagger \partial^\mu \phi$. We can fix this by introducing a new covariant derivative

$$D_\mu \phi = (\partial_\mu + i\frac{g}{2}\boldsymbol{\tau} \cdot \boldsymbol{A}_\mu)\phi. \tag{9.33}$$

Here A^i_μ are three new fields. If we can find a way to have this transform as

$$D'_\mu \phi' = U(x)D_\mu \phi, \tag{9.34}$$

then the Lagrangian

$$\mathcal{L} = (D_\mu \phi)^\dagger (D_\mu \phi) - m^2 \phi^\dagger \phi \tag{9.35}$$

will be invariant. To make this new derivative covariant as in equation (9.33) requires that A^i_μ transform as well, with

$$\boldsymbol{\tau} \cdot \boldsymbol{A}'_\mu = U^\dagger \boldsymbol{\tau} \cdot \boldsymbol{A}_\mu U - \frac{2i}{g} U^\dagger \partial_\mu U. \tag{9.36}$$

The first term just mixes the various A^i_μ fields in a particular way. It is the second term that is interesting as it is an additive term, a gauge transformation.

The photon has its own Lagrangian, $-\frac{1}{4}F_{\mu\nu}F^{\mu\nu}$, which also is gauge invariant. For these new gauge fields, we can make a similar construction. Note that the transformation of the covariant derivative in equation (9.34) is equivalent to $D'_\mu = U D_\mu U^\dagger$, so we can make the following construction using the commutator

$$[D_\mu, D_\nu]\phi = i\frac{g}{2}\,\tau^i F^i_{\mu\nu}\phi\,, \tag{9.37}$$

which would then produce the transformation property

$$\tau^i F^i_{\mu\nu}{}' = U\tau^i F^i_{\mu\nu} U^\dagger\,. \tag{9.38}$$

A little algebra using the properties of the Pauli matrices yields

$$F^i_{\mu\nu} = \partial_\mu A^i_\nu - \partial_\nu A^i_\mu + ig\epsilon^{ijk}A^i_\mu A^j_\nu\,, \tag{9.39}$$

which is similar to the photon field strength tensor aside from the addition of the last interaction term. Similarly the Lagrangian for this field can be constructed

$$\mathcal{L} = -\frac{1}{8}\,\mathrm{Tr}[\tau^i F^i_{\mu\nu}\tau^j F^{j\mu\nu}] = -\frac{1}{4}\,F^i_{\mu\nu}F^{i\mu\nu}\,, \tag{9.40}$$

where gauge invariance is most apparent in the first form. If we take the free field limit by setting the coupling constant $g = 0$, this looks just like the photon Lagrangian, so to a first approximation we may think of these particles as similar to photons. However, in contrast to photons, they interact with themselves, as evidenced by the extra terms in the Lagrangian.

These fields, and generalizations to larger symmetries, are referred to as *Yang-Mills fields* after the theorists who first constructed this Lagrangian.[2] They are the fields of the weak gauge bosons of the electro-weak sector of the Standard Model, and they also give (with a larger symmetry group) the gluons of the strong interactions sector, Quantum Chromodynamics.

If one gets deeper into the quantization procedure, one finds a difference from photons. Within classical electrodynamics, we know how to use the gauge freedom to find the two propagating polarizations of electromagnetic waves, which become the two photon states. This reduces the four components of A_μ down to two. Even in quantum loop diagrams, it is only these two that contribute. However, for the Yang-Mills fields this does not always happen. In most choices of gauge there are unphysical components of the fields in loop diagrams. In these cases, the unphysical components are removed by a trick of adding in extra fields that cancel the unphysical degrees of freedom. These are referred to as Feynman-DeWitt-Faddeev-Popov ghosts. Their construction is simplest in a path integral framework.

---

[2]C. N. Yang and R. L. Mills, "Conservation of Isotopic Spin and Isotopic Gauge Invariance," *Physical Review*, 1954, 96:191–195.

Gravitons have a construction that is somewhat similar to Yang-Mills fields although the relevant fields carry two Lorentz indices instead of one. The symmetry here is that of general covariance: the ability to make a change in coordinates independently at each point in spacetime.[3] Again there are covariant derivatives to be defined, and the equivalent of the field strength tensors are called curvature tensors. The quantization was carried out by Feynman and DeWitt in the 1960s. The result is not a renormalizable field theory, but it does still work as a quantum field theory using the rules of effective field theory.

9.2 Advanced techniques

We have described simple amplitudes treated at tree level, with a brief look at one-loop quantum corrections. These amplitudes can actually go far in many applications, but in the end the quantum world is more complicated. At the very least, higher orders in perturbation theory are often needed. Already at two loops, calculations are seriously difficult. In principle, all of perturbation theory is well-defined based on the Feynman rules. However, the techniques required at higher loop order become increasingly demanding. The calculation of the gyromagnetic ratio for the electron, which we quoted in chapter 1, has been carried out to four loops and that of the muon to five loops. These are tremendous accomplishments.

There are also advanced techniques that are not just the calculation of momentum-space matrix elements. For example, corrections to the Lagrangian can be discussed without forming matrix elements. We gave an example of this in section 7.7, on effective field theory, when we calculated the effect of the heavy $\bar{\sigma}$ field and gave the result as a new term in the Lagrangian involving the massless χ fields. This is a simple application of the *background field method*. The technique can be extended to include corrections to the action from the quantum corrections, even in cases where we are not discussing two separate fields. With a single field, we can imagine a separation of a field into a background value and a quantum correction, $\phi = \bar{\phi} + \phi_q$, where $\bar{\phi}$ is the background field. Expanding the Lagrangian to second order in the quantum field, we will have $\mathcal{L} = \mathcal{L}(\bar{\phi}) - \phi_q \, \mathcal{O}(\bar{\phi}) \, \phi_q$, where $\mathcal{O}(\bar{\phi})$ is some operator containing derivatives and the background field. The linear term in the expansion vanishes by the equations of motion. Advanced treatments of path integrals can then be used to show that

$$\int [d\phi_q] e^{-i \int d^4 x \phi_q \mathcal{O}(\bar{\phi}) \phi_q} = [\det \mathcal{O}(\bar{\phi})]^{-\frac{1}{2}} = e^{-\frac{1}{2} \int d^4 x \langle x | \log \mathcal{O}(\bar{\phi}) | x \rangle} , \qquad (9.41)$$

where the determinant is taken over all spacetime points and the final form comes from the mathematical identity $\det M = \exp(\mathrm{Tr} \log M)$. The final form can be evaluated as a new term in the action. We show you this scary formula not because we expect you to understand it at this stage but to illustrate that there is always

---

[3] As we have seen for the Yang-Mills theory, the gauge bosons A^i_μ carry two indices—one spacetime index μ and one index i associated to the field redefinition. Because in the case of General Relativity the field redefinition is given by coordinate transformations, whose parameters are the coordinates x^μ, the index i is replaced by a spacetime index. This is why the graviton carries two spacetime indices.

more Quantum Field Theory to learn. This last comment applies to many other advanced techniques also: they initially appear strange but with some effort are understandable.

9.3 Anomalies

Sometimes there are surprises that are not obvious outcomes of the Feynman rules. The existence of *anomalies* is one of these. An anomaly is said to occur when a symmetry of the Lagrangian is not a symmetry of the quantum theory. Noether's theorem is used to identify symmetries; specifically, these are symmetries of the Lagrangian that result in conserved currents. However, occasionally Noether's theorem misleads us.

The argument is easiest to state in the path integral context. The full quantum theory is the path integral and not just the Lagrangian. When a symmetry is anomalous, the Lagrangian is invariant but the full path integral is not. Symmetry transformations are transformations of the fields, and in some cases, the path integral over the fields has a nontrivial Jacobian determinant in the measure of the path integral when the fields are transformed. In the case of canonical quantization, the identification of an anomaly is generally found in treating loop diagrams and emerges from the fact that there is no regularization of the divergent diagrams that can be made consistent with the symmmetry.

One example of an anomaly is associated with scale transformations. A massless theory with no dimensionful couplings carries no Lagrangian parameters setting a scale, in other words, the theory does not contain a natural "unit of length." In this case, Noether's theorem can lead to a definition of a scale symmetry of the Lagrangian under a rescaling of fields and coordinates. However, we do not see such a symmetry in Nature because of a scale anomaly. While the analysis can be subtle, the basic physics of this comes from the need to choose an energy scale for the measurement of the coupling constant.

9.4 Many body field theory

We opened this book with an example—phonons—where a quantum field emerges from the interactions of a large number of particles. In section 6.3, we treated symmetry breaking in a field theory analogy of Ginzburg-Landau theories. There are many such examples where Quantum Field Theory is applicable to many body physics, both when the fields are emergent like the phonon or when they are part of the fundamental description of the system.

As an example, we can illustrate some techniques using the Ising model. This is a simple model for spin interactions with spins located on a lattice (which we will imagine is in three spatial dimensions), with the Hamiltonian

$$H = -J \sum_{\text{nn}(ij)} \sigma_i \sigma_j , \tag{9.42}$$

where nn(ij) denotes nearest neighbor locations. For the Ising model we take $\sigma_i = \pm 1$, mimicking the z-component of spin. The choice $J > 0$ is the ferromagnetic analog, with the spins energetically preferring to be aligned. The physics is described by a partition function, $\mathcal{Z}$, that involves the sum over all possible configurations of the spin

$$\mathcal{Z} = \sum_{\{\sigma\}} \exp\left\{-\frac{E[\sigma]}{T}\right\} . \tag{9.43}$$

At high temperatures we know that the spins will be randomly arranged and the average magnetization will be 0. As the temperature decreases, neighboring spins will be more likely to be aligned. However, there is a symmetry when you reverse the sign of all the spins, so that the overall Hamiltonian by itself does not have a preference on whether the spins are aligned and positive or aligned and negative.

Kadanoff and Wilson[4] suggested that it is useful to describe the average magnetization over *blocks* of spin on some spatial size $\sim L$ that is much larger than the lattice spacing, but much smaller than the size of the system. This can be described by a magnetization field

$$M(\mathbf{x}) = \frac{1}{N_L} \sum_{|\mathbf{x}_i - \mathbf{x}| < L} \sigma(\mathbf{x}_i) , \tag{9.44}$$

where N_L is the number of sites within this region, and where the coordinate $\mathbf{x}$, initially discrete, can be treated as a continuous variable. Each configuration of spins defined over the whole lattice then maps into a particular value of $M(\mathbf{x})$. We assume that we can capture the energy of these configurations with an approximate energy functional depending on the magnetization field $E[M(\mathbf{x})]$. The partition function then becomes a sum over all the possible magnetizations

$$\mathcal{Z} \sim \int [dM(\mathbf{x})]\, e^{-\frac{E[M(\mathbf{x})]}{T}} , \tag{9.45}$$

in the same sense that we define a path integral in field theory.

In looking for an approximate energy functional we can write out some simple expectations. We expect this energy to be obtained from an integration over all the positions on the lattice. It should be symmetric under $M \to -M$, as the Ising model has this symmetry. Rapid wiggles in the magnetization field should increase the energy. There should also be a dependence on the magnitude of the magnetization itself, not only on its gradients. These lead us to an approximate Ginzburg-Landau form

$$E[M(\mathbf{x})] = \int d^3x \left(\frac{k}{2}\, |\boldsymbol{\nabla} M(\mathbf{x})|^2 + a\, M^2(\mathbf{x}) + b\, M^4(\mathbf{x})\right) + \dots , \tag{9.46}$$

where the constants k and b must be positive to make the energy bounded from below, and all the constants k, a, and b will generally depend on the temperature.

---

[4]L. P. Kadanoff, "Scaling Laws for Ising Models Near $T(c)$," *Physics Physique Fizika*, 1966, 2:263–272; K. G. Wilson, "Renormalization Group and Critical Phenomena. 1. Renormalization Group and the Kadanoff Scaling Picture," *Physical Review B*, 1971, 4:3174–3183.

At this stage, it should be clear that a redefinition of the field $\sqrt{k}M(\mathbf{x}) \to \phi(\mathbf{x})$, followed by suitable parameter redefinitions, turns this into the $\lambda\phi^4$ theory that we have used heavily in this book:

$$E = \int d^3x \left(\frac{1}{2}|\nabla\phi(\mathbf{x})|^2 + \frac{\mu^2}{2}\phi^2(\mathbf{x}) + \frac{\lambda}{4!}\phi^4(\mathbf{x}) \right). \tag{9.47}$$

At the classical level, an ansatz for the coefficient μ^2 of the form $\mu^2 = c\,(T - T_c)$, with c a positive constant, would signal a phase change from a ground state with $\phi = 0$ above T_c to $\phi \neq 0$ below T_c, as we described previously in section 6.3. The appropriate blocking and the corresponding parameters can depend on the temperature, and in practice renormalization group techniques (see section 7.5) are used to describe the flow of these parameters when we change the temperature. By this construction, the many body Ising model has been turned into a field theory. Quantum field–theoretical diagrammatic techniques turn out to be extremely useful in evaluating the physics of a wide range of many body theories.

9.5 Nonperturbative physics

One can very rarely obtain exact analytic results in Quantum Field Theory. In this book, we have described the perturbative expansion of matrix elements in powers of a coupling constant. In many settings, this is the dominant approximation scheme for calculating observables. It is useful when the expansion parameter is small. However, it is not the complete answer. There are indications that even the perturbative expansion is only an *asymptotic expansion*. This implies that the power series appears to be convergent at low orders in the expansion but diverges at higher orders because of a factorial growth to the number of contributions. However, as long as the expansion is truncated before the factorially large terms kick in, we obtain a result that is very close to the "actual" one. The fact that the perturbative expansion is formally divergent should not make you doubt of its validity for the leading terms in the expansion: such a divergence is understood to be an artifact of the expansion itself, and there are ways the series can be resummed to give a finite result, at least in principle.

Moreover, we often need to treat systems where the perturbative expansion does not capture all of the key physics. This is because there are some phenomena that are inherently nonperturbative. In ordinary quantum mechanics, tunneling is a phenomenon that is not treatable in an expansion in the interaction potential. In the Wentzel-Kramers-Brillouin approximation, the tunnelling amplitude,

$$e^{-\frac{1}{\hbar}\int \sqrt{2m(V-E)}}, \tag{9.48}$$

cannot be sensibly expanded in powers of the potential. Likewise, Quantum Field Theory features transitions between field configurations that are not calculable perturbatively. There are field configurations that are similar to the "under the barrier" wavefunctions of the Wentzel-Kramers-Brillouin method. In some settings, these are referred to as *instantons* for historical reasons. Typically the tunneling amplitude is of the order of

$$e^{-\frac{c}{g^2}}, \qquad (9.49)$$

where g is the coupling constant and c is some pure number depending on the normalization chosen for g. The phenomenon can involve the tunnelling under a potential barrier $V(\phi)$ that is classically forbidden or could describe changes in the ground state properties. You can immediately see why these effects cannot be treated in perturbation theory by trying to expand the amplitude in equation (9.49) in powers of g.

Another setting where perturbation theory fails is when the coupling constants are large. Expansions do not make sense when each order in the expansion is larger than the previous ones. It is then natural to think that we could then turn to computers to calculate what we can no longer do analytically. One clear first step is to approximate space and time by making them discrete. This reduces the continuous infinity of spacetime points to a discretely infinite set, which can be made finite by working in a box of finite size. But that still leaves open the issue of what should be calculated on this spacetime lattice. We do not want to simply try to have the computer recreate the expansion in the coupling. Here the path integral representation is useful. The path integral is computed by integrating the field variables at each point in the (now discrete) spacetime, weighted by e^{iS}. This seems doable, but a couple of modifications are needed.

The fact that e^{iS} is just a phase makes the calculation difficult. Even when we performed analytical calculations, we had to add factors of $i\epsilon$ to make the simplest path integral converge. To make the integrand well behaved, we can analytically continue the time variable from t to $-i\tau$. This is consistent with the *Wick rotation* that we do when evaluating loop diagrams (see the appendix). It changes the Lorentz-invariant distance into a Euclidean one $-(\tau^2 + x^2 + y^2 + z^2)$. The kinetic energy terms also now carry a uniform sign

$$(\partial_t \phi)^2 - (\nabla \phi)^2 \rightarrow -[(\partial_\tau \phi)^2 + (\nabla \phi)^2]. \qquad (9.50)$$

Pulling out various overall signs, we arrive at a path integral with an exponentially damped weight e^{-S_E} and a positive-definite Euclidean action

$$S_E = \int d\tau \, d^3x \left[\frac{1}{2} \left[(\partial_\tau \phi)^2 + (\nabla \phi)^2 \right] + V(\phi) \right], \qquad (9.51)$$

which improves the convergence of the path integral.

In addition, for the computer to perform the integration over the fields for an action with interactions, one needs to adopt a statistical approach. This involves variants of the Monte Carlo methods, which were developed as a procedure that uses statistical sampling to obtain numerical results in complicated calculations. In the present case, the sampling needs to be done judiciously because most field configurations give a large Euclidean action, leading to an exponentially suppressed contribution to the path integral. The variety of methods to choose efficiently the most relevant configurations lie outside the scope of this book. But in the end the available methods try to generate well-chosen configurations of fields such that statistical averages of observables in these configurations approximate well the corresponding real-world quantities.

9.6 Bogolyubov coefficients

Yet another peculiar feature of Quantum Field Theory is that, when a field is quantized in the presence of a nontrivial background, typically a time-dependent one, quanta of that field can be created. Using suggestive but very nonrigorous language, the time-dependent background, even if spatially homogeneous, provides the energy that can allow the conversion of vacuum fluctuations of a field into actual physical particles. Remarkably, the rate of particle production in this case can be computed exactly in several cases (as long as we neglect the loss of energy of the background associated to the creation of particles) and typically turns out to have a nonperturbative dependence on the coupling constants.

For example, let us consider a field ϕ with a time-dependent mass $m(t)$ and assume for instance that the mass goes to m_i for $t \to -\infty$ and to m_f for $t \to +\infty$. Then, we cannot decompose the field as in equation (3.15), because such a decomposition does not satisfy the equation of motion with a time-dependent mass. Instead, we have to decompose the field as

$$\phi(t, \mathbf{x}) = \int \frac{d^3 k}{(2\pi)^3} \, [\hat{a}_{\mathbf{k}}^{(i)} \, \phi_k(t) \, e^{i\mathbf{k}\cdot\mathbf{x}} + \hat{a}_{\mathbf{k}}^{(i)\dagger} \, \phi_k(t)^* \, e^{-i\mathbf{k}\cdot\mathbf{x}}], \qquad (9.52)$$

where $\phi_k(t)$ satisfies the equation

$$\ddot{\phi}_k(t) + (k^2 + m^2(t))\phi_k(t) = 0, \qquad (9.53)$$

and if we want to identify the operators $\hat{a}_{\mathbf{k}}^{(i)\,(\dagger)}$ as creation/annihilation operators at early times, $t \to -\infty$, we must impose the initial condition

$$\phi_k(t \to -\infty) = \frac{e^{-i\omega_k^i t}}{\sqrt{2\,\omega_k^i}}, \qquad \omega_k^i = \sqrt{k^2 + m_i^2} \,. \qquad (9.54)$$

In particular, an observer that quantized the field at early times would define a vacuum $|0\rangle$ that is annihilated by the operator $\hat{a}_{\mathbf{k}}^{(i)}$, and will define the number operator as $\hat{N}_k^{(i)} = \hat{a}_{\mathbf{k}}^{(i)\dagger} \, \hat{a}_{\mathbf{k}}^{(i)}$. Because we will be working in Heisenberg representation, if the system initially has no particles, it will be in its vacuum $|0\rangle$ and will remain in that state for its entire evolution. Now, suppose that we have been able to solve the differential equation (9.53) exactly. (This is just an ordinary differential equation, and there are actually known solutions for several functional forms of $m(t)$.) Because at late times, $t \to +\infty$, $m(t) \to m_f$, the solution will be a linear combination of $e^{-i\omega_k^f t}$ and $e^{+i\omega_k^f t}$, with $\omega_k^f \equiv \sqrt{k^2 + m_f^2}$. We can thus write

$$\phi_k(t \to +\infty) = \frac{\alpha_k}{\sqrt{2\,\omega_k^f}} e^{-i\omega_k^f t} + \frac{\beta_k}{\sqrt{2\,\omega_k^f}} e^{+i\omega_k^f t}, \qquad (9.55)$$

where the *Bogolyubov coefficients* α_k and β_k can be computed by solving exactly equation (9.53) with the initial condition of equation (9.54).

Now, an observer that was born at late times will want to decompose the field into positive and negative frequency components as

$$\phi(t \to +\infty, \mathbf{x}) = \int \frac{d^3 k}{(2\pi)^3} \frac{1}{\sqrt{2\,\omega_k^f}} [\hat{a}_{\mathbf{k}}^{(f)} e^{-i\omega_k^f t + i\mathbf{k}\cdot\mathbf{x}} + \hat{a}_{\mathbf{k}}^{(f)\dagger} e^{i\omega_k^f t - i\mathbf{k}\cdot\mathbf{x}}], \qquad (9.56)$$

where, unless $\beta_k = 0$, the operator $\hat{a}_{\mathbf{k}}^{(f)}$ cannot coincide with $\hat{a}_{\mathbf{k}}^{(i)}$. Indeed, it is easy to see that

$$\hat{a}_{\mathbf{k}}^{(f)} = \alpha_k \,\hat{a}_{\mathbf{k}}^{(i)} + \beta_k^* \,\hat{a}_{-\mathbf{k}}^{(i)\,\dagger}. \qquad (9.57)$$

In particular, for an observer at late times, the particle number operator will be $\hat{N}_k^{(f)} = \hat{a}_{\mathbf{k}}^{(f)\dagger} \hat{a}_{\mathbf{k}}^{(f)}$, whose expectation value will read

$$\langle 0 | \hat{N}_k^{(f)} | 0 \rangle = |\beta_k|^2. \qquad (9.58)$$

We thus see that a nonvanishing value of β_k is interpreted as the generation of particles. Related uses of such Bogolyubov transformations appear in applications as diverse as superconductivity, Hawking radiation from black holes, and density perturbations in the Early Universe.

The first case in which such a situation was envisaged (even if it was studied using different techniques) was the possibility of creation of pairs of charged particles by an external uniform electric field, the so-called *Schwinger effect*. In the case of a scalar field with charge g, the equation of motion in the presence of an electric field of magnitude $E > 0$ directed along the positive z axis can be written in the form

$$\ddot{\phi}_k(t) + (k_x^2 + k_y^2 + (k_z - gEt)^2 + m^2)\phi_k(t) = 0, \qquad (9.59)$$

which is analogous to equation (9.53). In this case, we obtain a rate of pair production per unit time and per unit volume

$$\Gamma = \frac{(gE)^2}{(2\pi)^3} e^{-\frac{\pi m^2}{gE}}, \qquad (9.60)$$

which displays the nonperturbative behavior $e^{-1/g}$.

Chapter summary: Quantum Field Theory is not a finished subject. The present understanding of the subject is more advanced than it was thirty years ago. There are new applications and often we see new facets of Quantum Field Theory when we apply it in a novel setting. New calculational techniques are still being developed. And importantly, new concepts and insights are being recognized. The subject is alive and growing.

We hope that this short book has given you some insight into the ideas and techniques of Quantum Field Theory. If this is as far as you will go in the subject, hopefully you will have appreciated its elegance and picked up enough to follow its main applications. If you go on to use Quantum Field Theory professionally, you will surely grow in your appreciation of this subject.

APPENDIX

Calculating loop integrals

Perturbative quantum field theory involves loop integrals that capture the quantum corrections beyond tree-level effects. We might be tempted to not worry much because these are "just integrals." However, these are surprisingly subtle objects, and we need to evaluate them in order to extract all the physics. Here we give a very introductory treatment by calculating the simplest loop integral.

A.1 Basic techniques

In the treatment of scattering in $\lambda\phi^4$ theory, we encountered the loop integral

$$I(p^2) = -i \int \frac{d^4k}{(2\pi)^4} \frac{1}{(k-p)^2 - m^2 + i\epsilon} \frac{1}{k^2 - m^2 + i\epsilon}. \qquad (A.1)$$

Let us calculate this.

The denominator of this integral has poles when the propagators get close to the mass shell. A direct evaluation using the residues with these poles becomes complicated. However, these can be consolidated by using an identity

$$\frac{1}{ab} = \int_0^1 \frac{dx}{[ax + b(1-x)]^2}, \qquad (A.2)$$

such that

$$I(p^2) = -i \int_0^1 dx \int \frac{d^4k}{(2\pi)^4} \frac{1}{[k^2 - 2xk \cdot p + xp^2 - m^2 + i\epsilon]^2}. \qquad (A.3)$$

This is often referred to as a *Feynman parameterization*, and x is the Feynman parameter. The formula can be further simplified by defining $\ell_\mu = k_\mu - xp_\mu$ such that $\ell^2 = k^2 - 2xk \cdot p + x^2p^2$. Because $\int d^4k = \int d^4\ell$, we have

$$I(p^2) = -i \int_0^1 dx \int \frac{d^4\ell}{(2\pi)^4} \frac{1}{[\ell^2 + x(1-x)p^2 - m^2 + i\epsilon]^2}. \qquad (A.4)$$

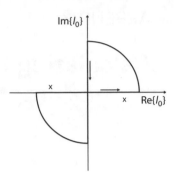

Figure A.1. Poles (marked by an *x*) and contour for the Wick rotation.

This looks almost like a spherically symmetric integral, but it is not because the Minkowski structure of $\ell^2 = \ell_0^2 - \boldsymbol{\ell}^2$ carries both possible signs.

The next step is the most subtle. We can turn this into a truly spherically symmetric integral using the trick of changing the ℓ_0 integration from the real axis to the imaginary axis, a process known as *Wick rotation*. To do this properly we need to make use of the $i\epsilon$ factor, which so far has played no role. If we let $a^2 = m^2 - x(1-x)\,p^2$, we see that there can be a pole in the denominator when $\ell_0^2 = (\boldsymbol{\ell}^2 + a^2 - i\epsilon)$ or $\ell_0 = \pm\sqrt{\boldsymbol{\ell}^2 + a^2 - i\epsilon}$. Letting ℓ_0 be a complex variable, we see that this falls below the real axis when ℓ_0 is positive, and above the real axis when it is negative, as in figure A.1. The easiest way to use contour integration is to chose a contour, like the one shown in figure A.1, that does not enclose any pole. The contour at infinity vanishes and so the integration over the imaginary ℓ_0 axis is equal to that over the real axis,

$$\int_{-\infty}^{\infty} d\ell_0\, f(\ell_0) = \int_{-i\infty}^{+i\infty} d\ell_0\, f(\ell_0) = i \int_{-\infty}^{+\infty} d\ell_4\, f(i\ell_4)\,, \tag{A.5}$$

where in the last form we have defined $\ell_0 = i\ell_4$ when along the imaginary axis. With $d^4\ell_E = d\ell_1\, d\ell_2\, d\ell_3\, d\ell_4$, we then have the four-dimensional spherically symmetric form

$$I(p^2) = \int_0^1 dx \int \frac{d^4\ell_E}{(2\pi)^4} \frac{1}{(\ell_E^2 + a^2 - i\epsilon)^2}\,. \tag{A.6}$$

At this stage the denominator is well behaved as long as a^2 is positive. This is not always the case, as when $p^2 > 4\,m^2$ we can see that $a^2 = m^2 - x(1-x)\,p^2$ can become negative at some values of x. But we will evaluate the integral first assuming $a^2 > 0$ and then deal with the negative case using the $i\epsilon$ prescription.

The four-dimensional Euclidean integral would be relatively easy to deal with if it were not for the next complication: the integral is actually divergent! This can be seen by counting powers of ℓ_E in the numerator and denominator, with the counting revealing that the integral goes as $\int d\ell_E/\ell_E$ at large momentum. In chapter 7, we described how this divergence disappears from physical observables, so this is not a disaster. But it does mean that we need to be careful in handling a divergent integral. In practice, this means employing some mechanism to make the integral into a finite one, that is, to *regularize* it. The mechanism is then removed later in the full calculation after the physics no longer is sensitive to it.

There are multiple regularization methods, and here we display two: *cutoff regularization* and *dimensional regularization*. The use of a cutoff is intuitively simple: just stop integration at a high energy scale, generally called Λ, $|\ell_E| < \Lambda$. In many cases this is fully acceptable. However in some theories, such as gauge theories, where there are symmetry reasons for having particles with no mass parameters, this can artificially introduce a new mass scale and cause troubles with the symmetry if we are not careful. In some cases the use of a momentum cutoff can also lead to problems with translation or Lorentz symmetry. These will not be an issue for us here. Dimensional regularization evaluates the loop integral in a different number of dimensions, $d < 4$, where it is finite, and then continues smoothly back to $d = 4$. The divergence appears as a factor of $1/(4 - d)$. This method is less intuitive, but it has the important feature that it manifestly preserves symmetries at all stages. Real physical quantities are ultimately independent of the regularization method, when properly applied.

To complete the integration, we first compute the angular part, defined in d dimensions by

$$\int d^d \ell_E = \int_0^\infty d\ell \, \ell^{d-1} \, d\Omega_d \,, \qquad (A.7)$$

where

$$\int d\Omega_d = \int_0^{2\pi} d\phi \int_0^\pi \sin\theta_1 d\theta_1 \int_0^\pi \sin^2\theta_2 d\theta_2 \dots \int_0^\pi \sin^{(d-2)}\theta_{(d-2)} d\theta_{(d-2)}$$

$$= \frac{2\pi^{d/2}}{\Gamma(d/2)} \,, \qquad (A.8)$$

with $\Gamma(x)$ being the Gamma function. This reproduces the usual results of 2π, and 4π in two and three dimensions, respectively, and yields $2\pi^2$ in four dimensions. However, expressed in this way, the result can be continued to noninteger, and even complex, dimensions. As we said, it is not very intuitive but it works.

For cutoff regularization, we stay in four dimensions and find

$$I(p^2) = \frac{2\pi^2}{(2\pi)^4} \int_0^1 dx \int_0^\Lambda d\ell \, \ell^3 \frac{1}{[\ell^2 + a^2]^2}$$

$$= \frac{1}{16\pi^2} \int_0^1 dx \left[\log\left(\frac{\Lambda^2}{m^2 - x(1-x)\, p^2 - i\epsilon} \right) - 1 \right] + O(\Lambda^{-2}) \,, \qquad (A.9)$$

where in the final form we have restored the definitions of a^2 and $i\epsilon$.

In dimensional regularization, it is standard to add a parameter μ with mass-dimension one to the integral to keep it dimensionless:

$$I(p^2) = \frac{2\pi^{d/2}}{\Gamma(d/2)(2\pi)^d} \int_0^1 dx \, \mu^{(4-d)} \int_0^\infty d\ell \, \ell^{d-1} \frac{1}{[\ell^2 + a^2]^2} \,. \qquad (A.10)$$

The parameter μ must disappear from observables when the regularization is removed. Consulting a good table of integrals we find

$$I(p^2) = \frac{\Gamma(2-d/2)}{(4\pi)^{d-2}} \int_0^1 dx \left[\frac{\mu^2}{m^2 - x(1-x)\, p^2 - i\epsilon} \right]^{2-d/2} . \tag{A.11}$$

Let us process this a bit more before discussing the physics. We are interested in the limit as $d \to 4$, so we define a parameter

$$\epsilon = \frac{4-d}{2}. \tag{A.12}$$

Please do not confuse this ϵ with the infinitesimal $i\epsilon$ in the propagator: both are conventionally denoted by the same symbol, but they can be readily told apart by context. Also be careful because some authors define $\epsilon = 4 - d$ without the factor of 2. In terms of this parameter, the divergent factor resides in

$$\Gamma(2-d/2) = \Gamma(\epsilon) = \frac{1}{\epsilon} - \gamma_E + \mathcal{O}(\epsilon), \tag{A.13}$$

where $\gamma_E = 0.57721...$ is the Euler constant. We can also make use of the identity

$$\frac{1}{\epsilon}X^\epsilon = \frac{1}{\epsilon}e^{\epsilon \log X} = \frac{1}{\epsilon} + \log X + \mathcal{O}(\epsilon). \tag{A.14}$$

These turn the dimensionally regularized integral into

$$I(p^2) = \frac{1}{16\pi^2} \left[\frac{1}{\epsilon} - \gamma_E + \log(4\pi) + \int_0^1 dx \, \log \left(\frac{\mu^2}{m^2 - p^2 x(1-x) - i\epsilon} \right) \right]. \tag{A.15}$$

Now to the physics. We saw in chapter 7 that when applied in $\lambda\phi^4$ theory, some portion of this integral is included in the measured value of the parameter λ in the process of renormalization. In one scheme, this was the integral $I(0)$. This is

$$I(0) = \frac{1}{16\pi^2} \int_0^1 dx \left[\log \left(\frac{\Lambda^2}{m^2} \right) - 1 \right], \tag{A.16}$$

using the cutoff and

$$I(0) = \frac{1}{16\pi^2} \left[\frac{1}{\epsilon} - \gamma_E + \log(4\pi) + \log \left(\frac{\mu^2}{m^2} \right) \right] \tag{A.17}$$

in dimensional regularization. The divergence is contained here and, once the renormalization program works its magic, it will disappear in all processes. The residual factor,

$$\Delta I(p^2) = I(p^2) - I(0) = \frac{1}{16\pi^2} \int_0^1 dx \, \log \left(\frac{m^2}{m^2 - p^2 x(1-x) - i\epsilon} \right), \qquad (A.18)$$

is the same in both schemes. The p^2 dependence here contains the real physics as it influences the momentum behavior of amplitudes.

Finally, we can get a result in closed form by evaluating the integral in dx. In cases where the factor inside the logarithm changes sign, the $i\epsilon$ determines the proper sheet for the logarithm (see section A.3). The results are

$$\Delta I(p^2) = \begin{cases} \frac{1}{16\pi^2} \left[\sigma \log \frac{\sigma-1}{\sigma+1} + 2 \right], & (p^2 < 0), \\[2mm] \frac{1}{16\pi^2} \left[-2\bar{\sigma} \, \tan^{-1}(1/\bar{\sigma}) + 2 \right], & (0 \le p^2 \le 4m^2), \\[2mm] \frac{1}{16\pi^2} \left[\sigma \log \frac{1-\sigma}{1+\sigma} + 2 + i\pi \, \sigma \right], & (p^2 > 4m^2), \end{cases} \qquad (A.19)$$

with

$$\sigma \equiv \sqrt{1 - \frac{4m^2}{p^2}}, \qquad \bar{\sigma} \equiv \sqrt{\frac{4m^2}{p^2} - 1}. \qquad (A.20)$$

A.2 Locality

The calculation of the previous section also highlights an effect that is useful in understanding effective field theories, which we studied in section 7.7: divergences in loops are equivalent to local effects. This effect is not immediately obvious because propagators describe the correlation of fields at different points, so that we would expect some nonlocal effects in the result. However, this comes from the finite piece of the loop, not the divergent piece. The physics of this division goes back the uncertainty principle. The ultraviolet divergence comes from high energy and hence appears as short distance physics. Long distance nonlocal physics comes from low energy, where there is no ultraviolet divergence.

We can see this mathematically by looking at the position space dependence of the loop integral of section A.1. We note that the momentum space loop is the Fourier transform of two position space propagators

$$I(p^2) = -i \int d^4x \, e^{ip \cdot x} \, D_F(x) \, D_F(-x). \qquad (A.21)$$

We can then Fourier transform the result back to position space

$$D_F(x) \, D_F(-x) = i \int \frac{d^4p}{(2\pi)^4} \, e^{-ip \cdot x} I(p^2), \qquad (A.22)$$

which can be broken up into two pieces

$$D_F(x)\,D_F(-x) = i \int \frac{d^4 p}{(2\pi)^4}\, e^{-ip\cdot x}\,[I(0) + (I(p^2) - I(0))]$$

$$= I(0)\,\delta^{(4)}(x) + \Delta D^2(x)\,. \tag{A.23}$$

This is our desired result. The divergence resides in the first term, which is local and proportional to $\delta^4(x)$. The residual $\Delta D^2(x)$ is finite and not easy to calculate, but it is demonstrably not local.

A.3 Unitarity

Let us briefly discuss the $i\pi$ in the third line of equation (A.19). This is more important than it looks, as it is connected to the unitarity of the S-matrix.

First, a bit about the mathematics. The logarithm is a multivalued function in the complex plane because of the property

$$\log(|R|e^{i\theta}) = \log|R| + \log e^{i\theta} = \log|R| + i\theta\,. \tag{A.24}$$

Even though $|R|e^{i\theta}$ is periodic in θ, with period 2π, the logarithm appears to get different values as $\theta \to \theta + 2\pi$. This ambiguity is usually dealt with by imposing that θ is constrained to live in a certain range, with the most common choice being $-\pi < \theta \leq \pi$. For negative arguments of the logarithm, $R < 0$, we are precisely along this "cut" on the complex plane, and this is where the $i\epsilon$ becomes important. For the purpose of our discussion here, it is sufficient to take p^2 very large, such that

$$\Delta I(p^2) = -\frac{1}{16\pi^2}\log\left(\frac{-p^2 - i\epsilon}{m^2}\right) + O(m^2/p^2)\,. \tag{A.25}$$

Thanks to the $i\epsilon$, then, we obtain the result

$$\log\left(\frac{-p^2 - i\epsilon}{m^2}\right) = \log\left(\frac{|p^2|}{m^2}\right) - i\pi\,\Theta(p^2)\,, \tag{A.26}$$

where Θ denotes the Heaviside step function.

The physics is much more interesting. Let us start with what appears to be a digression. The S-matrix describes the scattering of initial states to final states, with the "matrix" labels being a complete set of asymptotic on-shell states. The fact that probability is not lost in the scattering process means that any initial state scatters to the final states with unit probability. Because the final states are a complete set, this requirement can expressed as the unitarity of the S-matrix

$$S^\dagger S = 1\,. \tag{A.27}$$

Because we have defined the scattering matrix in the form

$$S = 1 + iT \tag{A.28}$$

this translates into

$$iT - iT^{\dagger} = -T^{\dagger}T. \tag{A.29}$$

As a matrix, the multiplication on the right side consists of the summation over a complete set of on-shell states.

Now go back to the $\phi + \phi \rightarrow \phi + \phi$ scattering process in which we encountered the loop integral of equation (A.1). The initial and final states both contain states with two quanta of ϕ, and we see that the amplitude must contain an imaginary part if we are to satisfy equation (A.29). To the extent that it is a one-channel process (i.e., neglecting possible intermediate states with four or more quanta of ϕ), the only on-shell intermediate state is the $\phi + \phi$ state itself. This is seen pictorially in figure 7.2(b) and is represented by the loop integral $I(s)$. We saw from the evaluation of the integral that this does pick up an imaginary part for $s > 4m^2$, which is the region where the intermediate state can be on-shell. In the integration over the Feynman parameter x, the imaginary part surfaces only for those values of x and p^2 where the on-shell condition is satisfied. The $i\epsilon$ in the particle propagator only becomes relevant as the momentum reaches the on-shell point.

This correspondence is true more generally than in this one reaction. Imaginary parts surface in Feynman integrals when on-shell intermediate states are possible. These imaginary parts are needed to satisfy the requirements of unitarity.

A.4 Passarino-Veltman reduction

Loop integrals are generally more complicated than the one that we evaluated in section A.1, as they can contain extra propagators in the denominator (often with different masses) or momentum factors in the numerator. Some of them are seriously difficult.

However, there is a procedure that is quite useful. The *Passarino-Veltman reduction theorem* states that all one-loop diagrams, no matter how many propagators nor how many factors in the numerator, can be expressed in terms of the scalar tadpole, bubble, triangle, and box diagrams. The phrase scalar implies that there are no momentum factors in the numerator. The remaining nomenclature refers to diagrams with one, two, three, or four propagators. In section A.1, we evaluated the scalar bubble diagram with equal masses.

Here is a simple example of how this works. Consider the following integral, involving three propagators and a numerator factor:

$$I_3(p_1, p_2) = \int \frac{d^4k}{(2\pi)^2} \frac{k \cdot p_1}{[(k - p_1)^2 - m_1^2] [(k - p_2)^2 - m_2^2] [k^2 - m_3^2]}. \tag{A.30}$$

This is simplified using the following identity for the numerator:

$$k \cdot p_1 = -\frac{1}{2} \left\{ [(k - p_1)^2 - m_1^2] - [k^2 - m_3^2] + [m_1^2 - m_3^2 - p_1^2] \right\} . \qquad (A.31)$$

Using this, the first two terms will cancel propagators, leading to scalar bubble diagrams. The numerator in the third term is independent of the loop momentum and can be pulled outside of the integral, leaving a scalar triangle diagram. The reduction of other integrals can be more complicated than this, but there are computer programs that can help.

Chapter summary: The calculation of loop integrals is technically challenging. As far as this book is concerned, this calculation is a mathematical task best left to the professionals. We have presented a taste of the process here. In reality, it is an interesting subject in itself, with ties to important physics such as causality and unitarity. If you intend to become a professional in Quantum Field Theory, it is an important topic for future mastery.

Bibliography

Alexander Altland and Ben Simon, *Condensed Matter Field Theory* (Cambridge University Press, Cambridge, 2010).

J. D. Bjorken and S. D. Drell, *Relativistic Quantum Fields* (McGraw-Hill, New York, 1965).

N. N. Bogoliubov and D. V. Shirkov, *Quantum Fields* (Addison-Wesley, Reading, MA, 1982).

C. P. Burgess, *Introduction to Effective Field Theory* (Cambridge University Press, Cambridge, 2021).

Ta-Pei Cheng and Ling-Fong Li, *Gauge Theories of Elementary Particle Physics* (Oxford University Press, Oxford, 1988).

John F. Donoghue, Eugene Golowich, and Barry R. Holstein, *Dynamics of the Standard Model*, second edition (Cambridge University Press, Cambridge, 2014).

John F. Donoghue, M. M. Ivanov, and A. Shkerin, *EPFL Lectures on General Relativity as a Quantum Field Theory* (arXiv:1702.00319 [hep-th]).

Eduardo Fradkin, *Field Theories of Condensed Matter Physics* (Cambridge University Press, Cambridge, 2013).

Eduardo Fradkin, *Quantum Field Theory: An Integrated Approach* (Princeton University Press, Princeton, 2021).

Barry R. Holstein, *Topics in Advanced Quantum Mechanics* (Addison Wesley, Reading, MA, 1992).

Claude Itzykson and Jean-Michel Drouffe, *Statistical Field Theory* (Cambridge University Press, Cambridge, 1991).

Claude Itzykson and Jean-Bernard Zuber, *Quantum Field Theory* (McGraw-Hill, New York, 1980).

M. Le Bellac, *Thermal Field Theory* (Cambridge University Press, Cambridge, 1996).

M. E. Peskin and D. V. Schroeder, *An Introduction to Quantum Field Theory* (CRC Press, Boca Raton, FL, 1995).

Alexey A. Petrov and Andrew E. Blechman, *Effective Field Theories* (World Scientific, Singapore, 2014).

Stefan Pokorski, *Gauge Field Theories*, second edition (Cambridge University Press, Cambridge, 2000).

P. Ramond, *Field Theory: A Modern Primer* (Benjamin/Cummings, Reading, MA, 1982). (First reprint: Frontiers in Physics 1981, 51:1; second reprint: *Frontiers in Physics* 1989, 74:1.)

M. D. Schwartz, *Quantum Field Theory and the Standard Model* (Cambridge University Press, Cambridge, 2013).

R. Shankar, *Quantum Field Theory and Condensed Matter: An Introduction* (Cambridge University Press, Cambridge, 2017).

Mark Srednicki, *Quantum Field Theory* (Cambridge University Press, Cambridge, 2007).

G. F. Sterman, *An Introduction to Quantum Field Theory* (Cambridge University Press, Cambridge, 1993).

Martinus Veltman, *Diagrammatica: The Path to Feynman Diagrams* (Cambridge University Press, Cambridge, 1994).

Steven Weinberg, *Quantum Theory of Fields*, volumes 1 to 3 (Cambridge University Press, Cambridge, 2013).

A. Zee, *Quantum Field Theory in a Nutshell* (Princeton University Press, Princeton, 2010).

Jean Zinn-Justin *Quantum Field Theory and Critical Phenomena* (Oxford University Press, Oxford, 2002).

Index